How Engineers Create the World
The Public Radio Commentaries of Bill Hammack

Also by Bill Hammack

Why Engineers Need to Grow a Long Tail • 2011

How Engineers Create the World
The Public Radio Commentaries of Bill Hammack
The Complete Collection

To Jay Pearce
Public Radio Impresario

Articulate Noise Books

Second Edition: September 2012

Hammack, William S.
    How engineers create the world: The public radio commentaries of Bill Hammack (the complete collection) / Bill Hammack - 1st edition (version 1.1)
    ISBN 978-0-983-96610-4 (pbk)
    ISBN 978-0-983-96611-1 (electronic)
    1. Engineering. 2. Engineering design. 3. Engineering – Psychological aspects. I. Title.

Thematic Table of Contents

Inventions & Inventors

Air Conditioning *105*   Better Mousetraps *391*   Blue LED *385*
Bob Kearns *389*   Bose Wave Radio *69*   Coffee Maker *27*
Color Film Chemistry *413*   Composting Toilets *89*   Concrete *113*
Contact Lens *63*   Cornstarch Packing Peanuts *73*   Cruise Control *185*
Directional Sound *181*   Duct Tape *263*   Elevators *13*   Gas Lighting *55*
Geiger Counters *187*   George Eastman & Kodak *293*   Glass *59*
Glo-sheet *209*   Glowing Hockey Puck *141*   Gore-Tex *153*   GPS *189*
Grain Elevators *29*   Hazel Bishop *287*   Jack Kilby *415*   Jell-O *215*
Lava Lamp *33*   Maclaren Strollers *279*
Margaret Knight & Paper bags *83*   Matches *65*   Mauve *125*
Moen Faucets *81*   Mood Rings *161*   Neon *299*   Nitinol *281*
Nylon *377*   O-Rings *17*   Phillips Screws *191*   Philo Farnsworth *291*
Photocopiers *7*   Pop Rocks Candy *127*
Pumpkin masters *35*   Razor *3*   RFID *363*   Saul Griffith *329*
Scotch tape dispenser *47*   Sliced Bread *277*   Spam *51*   Stay-on Tab *71*
Steadicam *21*   Superglue *99*   Super Soaker *61*   Surveillance Tags *165*
Swimsuits *133*   Television Remote Control *231*   Tupperware *203*
Typewriters *39*   Ultrasound imaging *53*   Vacuum Cleaners *23*
Vaseline *179*   Velcro *49*   Video games *37*   Volkswagen Beetle *275*
Wind-Up Radio *383*

Art, Literature & Music

The Beatles *137*   eBooks *45*   Fahrenheit 451 *339*   Hammond Organ *107*
Jekyll & Hyde *31*   Leonardo da Vinci *131*   Lord of the Rings *139*
Pompeii the Novel *323*   Project Gutenberg *43*   Saxophone *193*
Theremin *295*   TiVo *325*   Zildjian Cymbals *235*

Technology & Society

Accurate Throwing *311*   Anthrax *119*   The Calendar *333*
Cigarette Machine *109*   Color Film & Determinism *101*   Colton *233*
Container Ships *373*   Counting People *247*   Cryonics *285*   Diapers *205*
Education & Technology *19*   The Electrical Chair *301*
The Electronic Paper Trail of Terrorism *337*   Escapement *335*
Face Recognition Software *123*   Genetically Modified Food *11*
Gridlock *149*   HeLa Cells *259*   Housework *405*
Laundry Machines *261*   Machine Guns *15*   Thomas Midgley *207*
Numbers *129*   Numbers & Empire *211*   Potholes *253*   Risk *343*
SARS *255*   Science & the Declaration *91*   Technology & 911 *195*
Technology & Terrorism *117*   Voting & Paper *327*
Voting machines & gambling *375*   Waterless Urinals *419*

Computers & Information Technology

Adam Osborne *251*   Apple's Killer App *371*   Atomic Clock *313*
Cell Phones *79*   Claude Shannon *75*   Disc vs Disk *163*   Digital data *85*
Dvorak Keyboard *421*   Fiber Optics *269*   First E-mail Message *121*
Google *249*   IETF & Domain Names *95*   Iridium Satellites *199*
Landlines & the Five Nines *357*   LEO Computer *273*
Monocultures & viruses *289*   Moore's Law *403*   The Mouse *245*
MUZAK *219*   Open Source *409*   J.R. Pierce & Satellites *175*
Public Key Encryption *369*   Rolodex *197*   Silicon *237*   Spam *271*
Thomas Stockham *171*   Text Messages *349*   The Telegraph *297*
The Queen of Voice mail *155*   Voice Over IP & 911 *411*

Energy & Environment

Batteries in the Frig *157*   Blackout Reading *347*   Electric Cars *103*
Garbage *221*   Hydrogen-Powered Cars *241*   Local Power *407*
Los Angeles *351*   Nitrogen Fixation *213*   Partially Zero Emissions *317*
The Power Grid *283*   Proven Oil Reserves *315*   Recycling *393*
Wind Energy *267*

Toys, Sports, Travel & Entertainment

Baseball Stats *417*   Bicycles *111*   eBay *345*   Ferris Wheels *387*
Frisbee *217*   The Ice Hotel *147*   Lego Toy Company *319*   Legos *135*
Nautilus Machines *177*   Olympic Flame *145*
Olympics & Technology *365*   On-line shopping *41*   Ping Putter *77*
Roller Coasters *93*   SCUBA *265*   Slinky *167*   Terry Bicycles *173*
Tour de France *359*   Waiting in Line *223*

Food & Cooking

Champagne *307*   Clarence Birdseye *243*   Cooking a Turkey *305*
Corks *143*   Ice Cream *97*   Microwave ovens *303*

Flight

Airships *9*   Black Box *321*   Boeing vs. Airbus *341*
Concorde *257*   Jet Takeoff *25*   SpaceShipOne *355*
Stall Warning Detector *1*

Engineering Methods & Design

Bathtubs *201*   Betamax vs VHS *367*   Bolts *87*   Demolition *67*
Engineering Heritage *183*   Eulogy for Old Technologies *331*
Fluid Flow *229*   Fracture Mechanics *159*   Henry Dreyfuss *225*
Mass Manufacturing *5*   Nanotechnology *361*
New vs. Old Technology *57*   Packaging *381*   Plasmas *353*
Sears Tower *379*   Strobe Light Photography *169*   Swatch *227*
Technological Optimism *115*   Why a Chair? *239*

Stamps

Richard Feynman *401*   Josiah Willard Gibbs *395*
Barbara McClintock *397*   John von Neumann *399*

Alphabetical Table of Contents

Accurate Throwing *311*
Air Conditioning *105*
Airships *9*
Anthrax *119*
Apple's Killer App *371*
Atomic Clock *313*
Baseball Stats *417*
Bathtubs *201*
Batteries in the Frig *157*
Beatles *137*
Betamax vs VHS *367*
Better Mousetraps *391*
Bicycles *111*
Birdseye, Clarence *243*
Bishop, Hazel *287*
Black Box *321*
Blackout Reading *347*
Boeing vs. Airbus *341*
Bolts *87*
Bose Wave Radio *69*
Calendar *333*
Cell Phones *79*
Chairs *239*
Champagne *307*
Cigarette Machine *109*
Coffee Maker *27*
Color Film & Determinism *101*
Color Film Chemistry *413*
Colton *233*
Composting Toilets *89*
Concorde *257*
Concrete *113*
Contact Lens *63*
Container Ships *373*
Cooking a Turkey *305*

Corks *143*
Cornstarch Packing Peanuts *73*
Counting People *247*
Cruise Control *185*
Cryonics *285*
Demolition *67*
Diapers *205*
Digital data *85*
Directional Sound *181*
Disc vs Disk *163*
Dreyfuss, Henry *225*
Duct Tape *263*
Dvorak Keyboard *421*
Eastman, George & Kodak *293*
eBay *345*
eBooks *45*
Education & Technology *19*
Electric Cars *103*
Electrical Chair *301*
Electronic Paper Trail of Terrorism *337*
Elevators *13*
Engineering Heritage *183*
Escapement *335*
Eulogy for Old Technologies *331*
Face Recognition Software *123*
Fahrenheit 451 *339*
Farnsworth, Philo *291*
Ferris Wheels *387*
Feynman, Richard *401*
Fiber Optics *269*
First E-mail Message *121*
Flu (1918) *309*
Fluid Flow *229*
Fracture Mechanics *159*

Frisbee *217*
Garbage *221*
Gas Lighting *55*
Geiger Counters *187*
Genetically Modified Food *11*
Gibbs, Josiah Willard *395*
Glass *59*
Glo-sheet *209*
Glowing Hockey Puck *141*
Google *249*
Gore-Tex *153*
GPS *189*
Grain Elevators *29*
Gridlock *149*
Griffith, Saul *329*
Hammond Organ *107*
Head Skis *151*
HeLa Cells *259*
Housework *405*
Hydrogen-Powered Cars *241*
Ice Cream *97*
Ice Hotel *147*
IETF & Domain Names *95*
Iridium Satellites *199*
Jekyll & Hyde *31*
Jell-O *215*
Jet Takeoff *25*
Kearns, Bob *389*
Kilby, Jack *415*
Knight, Margaret *83*
Landlines & the Five Nines *357*
Laundry Machines *261*
Lava Lamp *33*
LED (blue) *385*
Lego Toy Company *319*
Legos *135*
LEO Computer *273*

Leonardo *131*
Local Power *407*
Lord of the Rings *139*
Los Angeles *351*
Machine Guns *15*
Maclaren Strollers *279*
Mass Manufacturing *5*
Matches *65*
Mauve *125*
McClintock, Barbara *397*
Microwave ovens *303*
Midgley, Thomas *207*
Moen Faucets *81*
Monocultures & viruses *289*
Mood Rings *161*
Moore's Law *403*
Mouse (computer) *245*
MUZAK *219*
Nanotechnology *361*
Nautilus Machines *177*
Neon *299*
Neumann, John von *399*
New vs. Old Technology *57*
Nitinol *281*
Nitrogen Fixation *213*
Numbers *129*
Numbers & Empire *211*
Nylon *377*
O-Rings *17*
Olympic Flame *145*
Olympics & Technology *365*
On-line shopping *41*
Open Source *409*
Osborne, Adam *251*
Packaging *381*
Partially Zero Emissions *317*
Phillips Screws *191*

Photocopiers 7
Pierce, J.R. & Satellites 175
Ping Putter 77
Plasmas 353
Pompeii the Novel 323
Pop Rocks Candy 127
Potholes 253
Power Grid 283
Project Gutenberg 43
Proven Oil Reserves 315
Public Key Encryption 369
Pumpkin masters 35
Razor 3
Recycling 393
RFID 363
Risk 343
Roller Coasters 93
Rolodex 197
SARS 255
Saxophone 193
Science & the Declaration 91
Scotch tape dispenser 47
SCUBA 265
Sears Tower 379
Shannon, Claude 75
Silicon 237
Sliced Bread 277
Slinky 167
SpaceShipOne 355
Spam (electronic) 271
Spam (meat) 51
Stall Warning Detector 1
Stay-on Tab 71
Steadicam 21

Stockham, Thomas 171
Strobe Light Photography 169
Super Soaker 61
Superglue 99
Surveillance Tags 165
Swatch 227
Swimsuits 133
Technological Optimism 115
Technology & 911 195
Technology & Terrorism 117
Telegraph 297
Television Remote Control 231
Terry Bicycles 173
Text Messages 349
Theremin 295
TiVo 325
Tour de France 359
Tupperware 203
Typewriters 39
Ultrasound imaging 53
Vacuum Cleaners 23
Vaseline 179
Velcro 49
Video games 37
Voice mail 155
Voice Over IP & 911 411
Volkswagen Beetle 275
Voting & Paper 327
Voting machines & gambling 375
Waiting in Line 223
Waterless Urinals 419
Wind Energy 267
Wind-Up Radio 383
Zildjian Cymbals 235

Chronological Table of Contents

Stall Warning Detector *1*

Razor *3*

Mass Manufacturing *5*

Photocopiers *7*

Airships *9*

Genetically Modified Food *11*

Elevators *13*

Machine Guns *15*

O-Rings *17*

Education & Technology *19*

Steadicam *21*

Vacuum Cleaners *23*

Jet Takeoff *25*

Coffee Maker *27*

Grain Elevators *29*

Jekyll & Hyde *31*

Lava Lamp *33*

Pumpkin masters *35*

Video games *37*

Typewriters *39*

On-line shopping *41*

Project Gutenberg *43*

eBooks *45*

Scotch tape dispenser *47*

Velcro *49*

Spam *51*

Ultrasound imaging *53*

Gas Lighting *55*

New vs. Old Technology *57*

Glass *59*

Super Soaker *61*

Contact Lens *63*

Matches *65*

Demolition *67*

Bose Wave Radio *69*

Stay-on Tab *71*

Cornstarch Packing Peanuts *73*

Claude Shannon *75*

Ping Putter *77*

Cell Phones *79*

Moen Faucets *81*

Margaret Knight & Paper bags *83*

Digital data *85*

Bolts *87*

Composting Toilets *89*

Science & the Declaration *91*

Roller Coasters *93*

IETF & Domain Names *95*

Ice Cream *97*

Superglue *99*

Color Film & Determinism *101*

Electric Cars *103*

Air Conditioning *105*

Hammond Organ *107*

Cigarette Machine *109*

Bicycles *111*

Concrete *113*

Technological Optimism *115*

Technology & Terrorism *117*

Anthrax *119*

First E-mail Message *121*

Face Recognition Software *123*

Mauve *125*

Pop Rocks Candy *127*

Numbers *129*

Leonardo da Vinci *131*

Swimsuits *133*

Legos *135*

The Beatles *137*

Lord of the Rings *139*

Glowing Hockey Puck *141*
Corks *143*
Olympic Flame *145*
The Ice Hotel *147*
Gridlock *149*
Head Skis *151*
Gore-Tex *153*
The Queen of Voice mail *155*
Batteries in the Frig *157*
Fracture Mechanics *159*
Mood Rings *161*
Disc vs Disk *163*
Surveillance Tags *165*
Slinky *167*
Strobe Light Photography *169*
Thomas Stockham *171*
Terry Bicycles *173*
J.R. Pierce & Satellites *175*
Nautilus Machines *177*
Vaseline *179*
Directional Sound *181*
Engineering Heritage *183*
Cruise Control *185*
Geiger Counters *187*
GPS *189*
Phillips Screws *191*
Saxophone *193*
Technology & 911 *195*
Rolodex *197*
Iridium Satellites *199*
Bathtubs *201*
Tupperware *203*
Diapers *205*
Thomas Midgley *207*
Glo-sheet *209*
Numbers & Empire *211*
Nitrogen Fixation *213*

Jell-O *215*
Frisbee *217*
MUZAK *219*
Garbage *221*
Waiting in Line *223*
Henry Dreyfuss *225*
Swatch *227*
Fluid Flow *229*
Television Remote Control *231*
Colton *233*
Zildjian Cymbals *235*
Silicon *237*
Why a Chair? *239*
Hydrogen-Powered Cars *241*
Clarence Birdseye *243*
The Mouse *245*
Counting People *247*
Google *249*
Adam Osborne *251*
Potholes *253*
SARS *255*
Concorde *257*
HeLa Cells *259*
Laundry Machines *261*
Duct Tape *263*
SCUBA *265*
Wind Energy *267*
Fiber Optics *269*
Spam *271*
LEO Computer *273*
Volkswagen Beetle *275*
Sliced Bread *277*
Maclaren Strollers *279*
Nitinol *281*
The Power Grid *283*
Cryonics *285*
Hazel Bishop *287*

Monocultures & viruses *289*
Philo Farnsworth *291*
George Eastman & Kodak *293*
Theremin *295*
The Telegraph *297*
Neon *299*
The Electrical Chair *301*
Microwave ovens *303*
Cooking a Turkey *305*
Champagne *307*
1918 Flu *309*
Accurate Throwing *311*
Atomic Clock *313*
Proven Oil Reserves *315*
Partially Zero Emissions *317*
Lego Toy Company *319*
Black Box *321*
Pompeii the Novel *323*
TiVo *325*
Voting & Paper *327*
Saul Griffith *329*
Eulogy for Old Technologies *331*
The Calendar *333*
Escapement *335*
The Electronic Paper Trail of
Terrorism *337*
Fahrenheit 451 *339*
Boeing vs. Airbus *341*
Risk *343*
eBay *345*
Blackout Reading *347*
Text Messages *349*
Los Angeles *351*
Plasmas *353*

SpaceShipOne *355*
Landlines & the Five Nines *357*
Tour de France *359*
Nanotechnology *361*
RFID *363*
Olympics & Technology *365*
Betamax vs VHS *367*
Public Key Encryption *369*
Apple's Killer App *371*
Container Ships *373*
Voting machines & gambling *375*
Nylon *377*
Sears Tower *379*
Packaging *381*
Wind-Up Radio *383*
Blue LED *385*
Ferris Wheels *387*
Bob Kearns *389*
Better Mousetraps *391*
Recycling *393*
Josiah Willard Gibbs *395*
Barbara McClintock *397*
John von Neumann *399*
Richard Feynman *401*
Moore's Law *403*
Housework *405*
Local Power *407*
Open Source *409*
Voice Over IP & 911 *411*
Color Film Chemistry *413*
Jack Kilby *415*
Baseball Stats *417*
Waterless Urinals *419*
Dvorak Keyboard *421*

Preface

Don't read this book straight through; just flip the book open at random. Let the pieces come at you by chance as they did for my public radio listeners. Each commentary is only two pages long, and in the form of the familiar or personal essay. A wonderful format that began with Montaigne's *Essais* and then given new life every century by great practitioners like Hazlitt, Orwell, and E.B. White. I can't hold a candle to their work: They wrote with elegance of the great issues of the day; I, instead, enjoy throughly the quotidian, the mass manufactured, the every day things made by engineers. I thought of updating the essays and considered ways to organize them—by date? by topic?—but rejected them all. They came at the public like life and they remain in this book like a conversation: Unorganized, slightly chaotic, yet with some sense that unity exists.

In re-reading the pieces to prepare them for publication I expected, of course, to see the unity created by a running commentary on engineering creativity, but I was surprised to see how much of my own life was revealed. In them it seems I am always in a restaurant, or on an airplane, browsing in a used bookstore, or getting something in the mail. My wife, father, and mother appear in them—even my shoes and underwear make an appearance! I'm surprised how much is revealed: Readers will learn that my wedding ring has my email address engraved on it, or that I have a fear of flying. I can see, too, the big events of my day: Terrorist attacking, industries appearing and disappearing, and political events occuring. That's partly, of course, a reflection that I wrote them for radio; a medium that prizes urgency and immediacy.

The commentaries began in 1996 when I decided to tell the public what engineers do. I explored various venues, but none really fit me. Plus, I didn't do a very good job selling my idea, usually I mumbled, or just didn't make the pitch right. So, all my attempts failed miserably. Then the perfect venue dawned on me: Public radio. I'd been a listener for years and knew it broadcast interesting, but thoughtful things.

So in the early summer of 1999 I sent the program director at my local public radio station—Jay Pearce at WILL-AM 580—a letter suggesting we meet. Also, I enclosed a tape of commentaries I'd made at home. This time, though, knowing in my soul that this would be the right place, I didn't want to fail.

So, I purchased a book on selling. I combed through every page of *How to Master the Art of Selling*, and like a true academic made nineteen pages of notes! Learning especially about the most important aspect: Closing the sale. I read, with gusto, about things very exotic to me: Methods called the "Porcupine Close" and the "Sharp Angle Close." Fully armed, I made an appointment with the program manager, Jay Pearce.

On the appointed day I went to his office, with a thick wad of notes on how to persuade the most reluctant buyer, prepared to "sell" at full tilt. Before I got out a word he said "I got your letter and your tape. Let's start on Tuesday." Flummoxed, I started to use my sales techniques to close the deal until I realized that, well, I'd *closed* the deal.

The next day I showed up at the studio and the commentary series began. For the first few the *Morning Edition* announcer said I would "return from time to time," but every Tuesday I just kept showing up with more to say, and within a month or two they said I'd be back *every* Tuesday.

From 1999 to 2006 I created over 250 radio commentaries. They began at my local station, but broadened to include a nice-sized region in the midwest, some appeared on public radio's business show *Marketplace*. And many appeared on Radio National Australia during Robyn William's wonderful *Science Show*. Some essays I wrote over a decade ago and so I thought of updating the pieces. I smiled when I saw a mention—in this age of *Wikipedia!*—a reference to the *Encyclopedia Britannica*. I cringed to see how I missed on eBooks. And I was surprised to see I had to tell listeners that "Google is no amateur operation." I decided, though, to leave the pieces alone: Either a deeper idea resonates, which no old-fashioned example will destroy; or an idea's power fades with time in which case no updating of details would bring it back to life.

I left the airwaves in September of 2006, turning my attention to new media. Still, though, I miss the purity and freedom of a radio commentary. What a wonderful medium.

Stall Warning Detector

L AST WEEK I flew to Pennsylvania to visit a friend. A simple thing, yet for me it can be a terrifying thing. As an engineer I marvel every time I fly. I admire the jet's craftsmanship and am amazed at how routine it is to now fly 200 plus miles per hour. Yet I also do a very "unengineering-like" thing: As the jet shoots into the sky my palms become sweaty and my heart beats swiftly. I have a fear of flying.

I admire jets because they're a pinnacle of modern engineering and my training makes me believe the statistics showing flying is safe, yet still I'm a bit on edge every time I fly. Because I must fly I've had to conquer this fear of flying. I've learned that for me at the root, like most fearful flyers, is control. I'm used to being in control and when I sit in this huge plane ready to shoot into the sky someone else is in control. To overcome this fear I've had to learned everything about airplanes—about what all its many parts do. I use this knowledge as a talisman to calm me as we approach take off. One of my favorites is a chunk of metal a few inches long and only an inch or so wide on the fuselage or the wing.

It's a detector which tells the pilot if he or she is about to lose lift and thus crash the plane. In my study of airplane anatomy I learned how this device came about. I recently talked to its inventor Leonard M. Greene. He told me he'd invented the stall warning detector at age nineteen. This was a story I had to know so I asked him if this were the start of his long career in aviation—sixty patents in the field. "No," he said, "I'd say this all started when I was about five years old." Greene explained that he spent his childhood in extreme poverty.

With no money for toys he used his imagination helped by a children's encyclopedia which he'd found in the trash. The encyclopedia dazzled him with its experiments and self-made toys. A cherished memory is of a lighted wagon powered by batteries scoured from the trash—he'd learned from the encyclopedia how to rejuvenate them in salt water. Greene's mother saw this talent and pushed him to skip every other grade until he entered college at thirteen. Toward the end of his college years, around age seventeen, he stumbled on a science fiction novel that gave him his lifetime ambition to be an inventor.

He can rattle off all its predictions: fluorescent lights, baseball played under the Astrodome, and credit cards. So when, at nineteen, he saw an airplane drop from the sky he drew on his childhood to make the first stall warning detector becoming the inventor he dreamed of after reading science fiction. And Greene keeps that spirit alive today by continuing to invent: supersonic aircraft, wind shear warning systems, sailboat keels, a three-dimensional chess game, and even a ski binding.

It's people like Leonard Green childhood genius that give me confidence when I climb into a jet and zoom into the sky.

Razor

EVERY MORNING I STARE in my bathroom mirror and use a technological marvel that cost three-quarters of a billion dollars to develop. It is, of course, my razor. I just bought a new one, and the great engineering novelty is that it has three blades, not two.

For a long time razors used only two blades, but manufacturers have quested for the holy grail of a three bladed razor because it cuts more hair, yet doesn't irritate the skin. Sounds simple, but it really is very high tech engineering. Here are the steps to develop this razor.

Step one: Understand shaving. First an engineer developed a high-powered microscope to magnify freshly shaved chins forty times. Peering through this microscope the engineer used a laser-guided device to measure how much each hair was cut by a razor stroke.

Step two: This engineer passed this info to a colleague who wrote down a set of equations—called finite elements—to model shaving in a computer. This computer model lets other engineers study shaving before designing a new razor. From this work they learned how a razor works: As a razor moves it makes the skin bulge, forcing hairs up and out. The blade catches a hair, pulls it up, and slices through it, after which the hair starts to retract. And, in a two bladed razor, the second blade catches the hair before it can retract fully, and cuts it again. Now from this basic understanding the engineers could see that a third blade would cut the hair even shorter—cut 40% more hair—but they learned from their computer that this third blade would get too close and tear the skin slightly. But since this three bladed razor was the holy grail of razors the engineers played with their computer to make a third blade work.

They learned that tipping the third blade at an angle to the other two blades will cut the hair, but not significantly tear the skin. This is progress, but along comes step three because a problem arises: How to pack all three blades into a compact razor. They try making the blades thinner, but they become too fragile, to the point where a hair will actually break or blunt the edge. So another engineer enters. This time a metallurgical engineer—a specialist in metal—who realized you could cover the steel blade with a thin layer of something super hard and make the metal blade stronger.

Step three: That something super hard is a thin layer of carbon that makes a diamond like coating on the blade. Well, now we have a razor with three blades that are stronger than steel and cut hair better than anything else in the world. What's the next step?

Step four: Make a razor that appeals to consumers. The marketing team decided to give this razor a neat finish: They asked the engineers to watch Arnold Schwarzenegger's movie *Terminator 2* and match the finish on the razor's handle to the movies' liquid-metal villain, giving it an almost mercury like finish.

Then the last step, *step five:* Get a group of engineers to make a world class factory to produce the razor. The heart of this factory is a Class 5000 clean room with an environment more pure than any surgery ward. Here white coated workers carefully coat the steel blades with the hard carbon film. All this—three quarters of billion dollars worth of stuff—to make a simple razor, that you can buy for just a few dollars.

Mass Manufacturing

M Y PARENTS TOOK my brother, sister, and I on the oddest family vacations: We visited factories across North American. My parents prized these tours, especially my father, which was odd because my mother was the scientist, not him.

He was a theater professor who didn't even know the rudiments of science and engineering. Yet factories fascinated him: He stood spellbound as a Canadian factory made wood boards at a tremendous rate, or a plant in California filled jillions of cans of corn a minute. And because he was a cereal lover he marveled at the huge, smelly vats at Kellogg's in Michigan.

I learned how much my father valued these tours when we stopped in Detroit and he tried to arrange a tour of an auto maker. I remember him walking back to the car, his body telegraphing his disappointment: You had to be eighteen to have a tour, and my siblings and I were no where near that age.

As a child I, too, marveled at mass production. Now, as a full grown engineer, I marvel even more because I now know how much clever engineering goes into mass producing something.

The main thing is to save material and time. An engineer must be obsessive in shaving every ounce from a product. Think of a metal soda can: Three hundred millions cans are made every day. If you use just a tiny bit of extra metal on each can that tiny bit turns into a lot when multiplied by three hundred million cans a day. This is the reason a soda can narrows near the top—the slight curve at the neck of the can. This "necking" isn't for aesthetics; it's to save money. The top uses a quarter of the aluminum needed to make a can; by narrowing the can's neck the top is made smaller. For every four hundredth of an ounce shaved off the top, the can maker saves twenty million dollars in metal.

So, the first step in mass producing something is to shave weight, and the second is to save time. In production time is money. And this is where real cleverness enters. Take a golf ball. Its center is a liquid encased in a mass of rubber thread—this is what gives the ball its bounce. Now picture yourself as an engineer who's been asked to make this golf ball. How would you quickly wrap a ball of liquid with tons of rubber thread? Bear in mind that moving liquid around is tricky. Think of carrying a cup of coffee upstairs, often you spill it, unlike something solid like a donut, which is easy to carry. How then do you make a golf ball quickly? You freeze the liquid center—making it solid—and quickly wrap it in rubber thread.

Also very clever is the production of plastic wrap—the stuff you cover food with. I'm sure you've been frustrated trying to get it off the roll; it sticks to everything. How, then, in a mass production line, do they make this stuff? They take molten plastic and blow a big bubble of smooth plastic wrap, then collapse this bubble to create a single sheet of wrap. The main problem they have is with flies crashing into the wrap and wrecking the bubble before they want it to collapse.

How do I know all this about plastic wrap? One day, long after our family vacations to factories had ceased, I visited my father. He now lived alone. He looked at me and said "Would you like to visit a factory?" He had a wistful look so I said yes. And we trotted off to a nearby chemical factory and sat, father and son, for a quiet hour or two watching plastic wrap being made.

Photocopiers

IN THE SOVIET UNION in the 1970s were machines so powerful and so dangerous that they were kept locked and guarded around the clock. The machine: A photocopier.

Its danger, of course, was the spread of information not controlled by the communist regime. The history of this dangerous machine fascinates me because it shows clearly innovation from the beginning of the technology —from its invention to its marketing.

The photocopier began with Chester Carlson in the late 1930s. Carlson became bored while working as an engineer for the telephone company. Thinking patents would be more exciting he eventually took a job with a Patent Attorney, and worked toward a law degree at night. Newly married and needing cash, he decided to—in his own words—"to do the world some good" and "to do some good for himself."

He returned to a fascination from his youth—graphic arts and chemistry—to dream of a small machine that fit in an office and made one copy per second. About his invention he later wrote:

"Things [like this] don't come to mind .... all of a sudden .... You have to get your inspiration from somewhere and usually you get it from reading."

Carlson read every patent he could searching for ideas. He learned of electrostatics: things sticking together because of static electricity. He fiddled with electrostatics and chemicals in his kitchen, but the smells irritated his neighbors, so he moved to an apartment in Astoria, Queens.

On a fateful day in October of 1938 he coated a metal plate with sulfur power. Took out a glass slide and wrote: the date—ten dash twenty-two dash 38 and the town "Astoria." He pulled down the

shade, rubbed the sulfur surface with a handkerchief to create a charge
—like shuffling your feet across the carpet so you can shock yourself
on the wall. He laid the glass slide with writing against the plate and
shined light on it. After removing the glass he spread a dark powder
onto the metal plate; then he blew it off, but it stuck where the letters
"10-22-38 Astoria" had been. He pressed the plate against a piece of
paper making the world's first photocopy. He formed a company—
eventually called Xerox—and by 1949 was selling copiers.

But here is where more innovation enters: Marketing innovation. A
product's introduction to the marketplace is not the end of innovation
but merely the end of the beginning. Now imagine trying to sell a
huge expensive and unproven machine. Xerox's goal was to get the
machine into a few markets and then grow these until they connected
into a national demand for photocopiers. What was the key to
creating these local markets?

They realized it was hard to get someone to buy a machine outright,
so they developed, in 1959, a pricing system for leasing. For a base
monthly fee you got a machine and then paid based on the number of
copies you made. The price was low enough, and the contract short
enough that even the most reluctant office manager would give it a
try. They bet that people would love their product and make copy
after copy. They were right. The result: From nearly zero copies in
1959 to 490 million in 1966.

In fact, photocopiers have became so central to our lives, and
consume so much time that today you can even buy a book of
exercises to do while at the photocopier!

Airships

ELIUM HEAD. You hear me correctly: "helium head." Helium like in a balloon and head like in your head. I'm sure you don't know what this means, but its my response to the great French Philosopher Jacques Ellul, a potent critic of technology, who wrote "technology is essentially independent of the human being, who finds himself naked and disarmed before it."

As a lowly engineer I'm hesitate to argue with a philosopher, but I don't think technology crushes us, instead it often reflects us. I suppose my response of "helium head" hasn't often been heard in philosophical circles, but it is very much to the point. To be a "helium head" means you've fallen in love with lighter-than-air travel—airships, blimps, dirigibles, or zeppelins. And in them you see an instant and easy solution to nearly all the world's transportation problems. The disease of helium-heady-iness proves that technology reflects us—and not the other way around, in spite of French philosophers.

Airships reached their greatest popularity in the 1920s and 30s. In that period airplanes were small, noisy, and slow; unlike airships that were grand and serene. I quote Britain's Minister for Air Travel. He said, in 1930,

> *"[t]ravellers will journey tranquilly in air liners to the earth's remotest parts ... cruise round the coasts of continents, ... [and] surmount lofty mountain ranges ...."*

But with the explosion in 1937 of the Hindenburg, the airship industry folded. Yet in every decade since their demise engineers have tried to revive airships. Why, in this age of jets?

Because airships can lift great weights cheaply; ferrying goods, for example, from ships at sea, thus getting rid of expensive harbors. And airships pollute less and are quieter than jets. Each new airship proposed has reflected these strengths, but also in each decade the proposals reflect the interests of the time.

In the 1950s and 60s—a time when nuclear bombs preoccupied our minds—an engineering professor came up with a nuclear powered airship.

In the 1970s the so-called "Third World" rose in our consciousness: In that era engineers designed a radial airship that took off like a jet, yet floated and lifted like an airship. The minister behind this failed project saw it as a vast warehouse in the sky from which he could bring all nations into the 20TH century by a single leap: no need for roads, railroads, airports, warehouses, or harbors.

Airship designs continue to reflect societies interest of the moment: In the 1980s Engineers proposed using airships and blimps to halt ozone depletion over the South Pole by hanging live electrical wires from the ships to zap ozone-eating chemicals.

And a former U.S. Secretary of State heads a company to use blimps as communications satellites. But the mighty airship may rise again because there is one airship company left: The Zeppelin company.

It didn't fold when the Hindenburg failed, instead it became a multi-billion dollar construction company. They are building the first new Zeppelin in decades. To quote the chief engineer: "In the near future, thousands may enjoy the most buoyant tourist experience of their lives, gliding serenely over scenic vistas in the cabin of a 21ST-century zeppelin." Is this a true revival? Or is it just an engineer afflicted by the disease of being a helium head—of seeing in airships, and technology more generally, the solution to all our society's ills?

We'll see, but I'd bet on helium-heady-ness.

Genetically Modified Food

RECENTLY MY WIFE and I ate at a local Mexican restaurant and while there I studied their machine for making tortilla bread—oddly enough it reminded me of genetically modified food. Now compared to messing with genes a bread machine seems mundane—who doesn't have one in their home?—but in the 19TH century this machine would have been as controversial as today's genetically modified food.

To someone like 19th century minister Sylvester Graham a bread machine would be a kind of heresy. To Graham the problem with bread was not technical but spiritual. Graham spoke in favor of hard mattresses, cold showers, and a special diet. His great concern was to restore man's contact with nature using flour, bread and chewing.

"Bread," Graham said,

> "should be baked in such a way that it will ... require and secure a full exercise of the teeth in mastication."

In simple words: bread must be chewy. Small wonder Ralph Waldo Emerson called him "the prophet of bran bread and pumpkins." Graham advocated stale bread to promote chewing, but when this didn't catch on, he pushed a recipe for a coarse wheat bread.

It was whole wheat bread with a dash of doctrine. He insisted it be made by hand at home because even if a commercial baker followed Graham's recipes, the baker lacked "the moral sensibility" to make healthful bread. Graham said the baker could not grasp "the importance ... of bread, in relation to the happiness and welfare of those who consume it." Such a moral sensibility is not, thought Graham, found in commercial bakers. And he was deeply concerned about what he called "artificial chemical agents" used by bakers—the yeasts and salt we use today—and he even implied they tossed "chalk, pipe clay and plaster of paris" into their breads to increase their weight and whiteness. Not surprisingly, a mob of angry bakers once attacked him.

Graham's influence is still with us today: He invented the Graham cracker to make it convenient to chew wheat grain anytime of the day, and one of his many followers—called "Grahamites"—pressed flour into thin sheets, then ground it into bite-sized pieces, and baked it until hard. Thus inventing cold breakfast cereal.

Today we regard this 19TH century minister Sylvester Graham as something of a crank, but in a sense he was right. He feared that processed soft white bread lacked all nutrition.

In 1974 a medical journal suggested that fiber could reduce all sorts of diseases, igniting a firestorm against white bread and a return to "natural" breads that were less processed. So, was Graham right? Maybe, but the coin is flipping again: Science now tells us that white bread does indeed contain fewer nutrients than whole wheat, but that those in the wheat breads cannot be digested as easily, which is fine in our nutritionally rich western world, but in poorer nations a diet purely of unprocessed grain can tip the balance from health to serious disease.

So what does this story of Reverend Sylvester Graham and his fear of white bread tell us about genetically modified food? Well, just that it isn't purely a technical question—the use of these foods will require a consensus reached by a community based on their norms, their religious views, and their needs—and as we debate scientists and engineers will weigh in, but will not necessarily have the final word.

Elevators

WHAT IS THE SAFEST way to travel? Its a method that every year moves forty-five billion passengers a distance of one and a half billion miles—and with only about fifteen deaths.

It is an elevator.

Now it's useful only, of course, for traveling up and down a building; although, according to insurance companies, its five times less hazardous that climbing steps. Elevator were not always this safe; in the early 1800s people used them only for moving freight—no reasonable person would ride on such a thing—it was not uncommon to hear of a heavily loaded elevator plummeting and killing everyone aboard.

The key innovation in making elevators appealing to people came from Elisha Graves Otis—a name still seen on millions of elevators. In 1853, at the Crystal Palace Exposition in New York, Otis, dressed in tails and a top hat, stepped into one of his elevators, filled with boxes and barrels, and was hoisted forty feet above the ground. When the elevator stopped Otis ordered a workman with an ax to cut the hoist cable supporting the elevator. The crowd watched in horror as the platform jerked and then ... nothing happened: Otis and his elevator platform stayed right where they were.

As the stunned crowd stared Otis took off his top hat, bowed, and made his sales point, he cried out "All safe, ladies and gentlemen, all safe." A new rope was attached and the elevator lowered safely to the ground. Otis called this his "safety elevator." He built a spring loaded lever into the its floor. The hoist cable kept the lever retracted, but if it were cut and tension lost the spring instantly thrust the lever into the shaft stopping the elevator's descent. Today this same basic device still keeps elevators from tumbling down their shafts.

In spite of this dramatic demonstration it took Otis three years to sell a passenger elevator. His first was installed in a five-story china and glass emporium on Broadway in New York. It whisked passengers up at forty feet a minute—an elevator this slow would take thirty minutes to reach the top of Chicago's Sears Tower, a trip made in less than a minute with today's elevators. Otis' elevator company succeeded, unlike his previous business ventures partly because it became crucial in accommodating the large urban growth of the last two decades of the 1800s.

Because of a huge influx of workers house prices in cities skyrocketed—there simply wasn't enough land to go around. Elevators made apartment buildings attractive, which in the past had been segregated by economic class: Wealthy families lived on the lower few floors so they didn't have to climb many stairs, while poor families were usually confined to the basement or the upper floors.

The elevator allowed taller buildings to be built and, in a sense, democratized apartment buildings allowing all floors to be equally attractive. Today, of course, elevators are not viewed as a force for democracy, or even as a mechanical miracle that saves lives. Like all successful technological innovation our interest in them has been reduced to social psychology; to a set of rules governing our behavior in elevators—conversations should be whispered, eyes focused forward toward the door, and strangers should keep a distance from one another—all because Elisha Graves Otis attached a spring and a lever to bottom of an elevator.

Machine Guns

YESTERDAY MY WIFE and I were sitting at our computers each sending out e-mail when, in mid-message, by wife stopped, looked at me and asked how do you spell Kalashnikov? I hadn't a clue how to spell the name of this Russian machine gun. I respected her privacy and didn't ask why she wanted this for an e-mail message, although if I'm missing, you know where to start.

This talk of Kalashnikovs got me to thinking about machine guns. They are, of course, a ghastly thing, but like all technology they reflect directly our own human nature, our prejudices and world views.

The idea of a machine gun is very old—a man named Palmer described one in 1663—but the first fully automatic gun, designed by Hiram Maxim, appeared in 1884. Maxim was traveling in Vienna, when a fellow American, told him "if you want to make a pile of money," he said, "invent something that will enable these Europeans to cut each other's throats with greater facility." This resonated with Maxim, who returned to London and spent two years designing a gun.

Maxim's gun used the gases of the bullet's explosion to move a piston under the barrel, which worked the bolt, expelled the cartilage and reloaded the gun. You would think such a weapon of death would be widely adopted, but Maxim found great resistance to his weapon.

In European most officers came from the land owning classes, and left behind by the Industrial Revolution they still thought in terms of the bayonet push and the calvary charge. They clung to the belief "of the centrality of man and the decisiveness of personal courage and individual endeavor"—after all you can't pin a medal on a gun.

Although these officers felt it uncivil to use machine guns in European battle, Maxim found he could play on their prejudices in the colonies. He sold many guns in Africa, which European soldiers and settlers used against unarmed natives. Oddly these colonial slaughters tainted the machine gun, making it even less palatable for European Warfare.

In fact fresh recruits to World War I once asked their commanding officer what to do with their newly issued machine guns. He said "take the damned things to a flank and hide them." Although in World War I each side began with only a few machine guns, these few showed its tremendous powder, and by the end of the war the machine gun was an essential tool.

Siegfried Sassoon, one of the great solider-poets of the War, captured this in an ode to a machine gun:

> *"To these I turn, in these I trust -*
> *Brother Lead and Sister Steel.*
> *To his blind power I make appeal,*
> *I guard her beauty clean from rust."*

Another great war poet, Robert Graves, captured how the machine gun changed war forever:

> *"And we recall the merry ways of guns*
> *Like a child, dandelions with a switch!*
> *Machine-guns rattle toy-like from a hill,*
> *Down in a row the brave tin-soldiers fall."*

So simple is war, says Graves, now even a child can do it. The machine gun ushered in, perhaps forever, an age of horrible mechanized war, making it clear that a solider could no longer depend on personal courage or strength for victory or even survival; machine guns destroyed forever the illusion of courage, hope, and a sense of the heroic possibilities in war.

O-Rings

THERE EXISTS A tiny device without which we wouldn't easily have air planes, automobiles, tractors, or air conditioners—just about anything with motion. In fact you use one every time you step on the brake in your car. Its costs fractions of a penny and has made headlines only once. It is the tiny, yet mighty o-ring.

You've likely heard about o-rings during the Space shuttle disaster of 1986, and I'm sure you've seen one, the name is self-descriptive—they are usually just a thin tube of rubber shaped into a ring. It seems so simple and obvious that it couldn't have been really invented—yet it isn't that obvious, there is a trick to making an o-ring work. Usually we think of such a world changing invention as coming from some inspirited young turk, but the o-ring came from a senior citizen, Niels Christensen.

Christensen, a danish immigrant, was an expert on brakes. He'd came to America at age 26 to be the leading draftsman for a Chicago engineering firm, but after a year or two the company reorganized and he lost his job. While unemployed he read of a major streetcar crash where the breaks failed. How, he wondered, could he improve the breaking system. At that time the break shoes were pressed against the wheels by the strength of the conductor, amplified by the electricity that ran the train. The problem: A sudden loss of electrical power and the breaks were out. Christensen realized there needed to be a way to store the energy and release it later—to do this the ingenious inventor used air.

Before the car started out Christensen used an electric motor to force air into a cylinder, which when released drove the break drums. Because the air was stored and released mechanically it didn't matter if the electricity shut off. How does this relate to the tiny o-ring?

Christensen needed to seal the compressed air cylinder, the seal he used was a cumbersome and tricky triple value; he didn't use an o-ring, but it put the problem of sealing foremost in his mind. Some forty years later in 1933 Christensen, now sixty-eight, was still working on sealing the fluid in breaks—this time though for cars.

He tried this time a simple rubber ring. He cut a groove into his piston, slipped the o-ring over it and pressurized the container; he found, as others before him did, that it failed. If he had been a younger man he wouldn't have had the insight and intuition to continue. Patiently Christensen changed the size of the groove, cutting new ones with slightly different dimensions. In time he found the magic to an o-ring: Make the groove one and half times the o-ring radius. The result was remarkable: "This packing ring", he wrote in his notebook, "has been tested" nearly three millions times and "has never leaked and is still tight."

It was so simple that no one believed it would work until finally two World War II Army Air Crops engineers used it to fix some leaking breaks on the landing gear of Northrop bomber. It worked like magic and was used in all military aircraft—and soon this simple, but ingenious o-ring seal appeared everywhere: fountain pens, soap dispensers, plumbing systems, hydraulic presses, automobile breaks, washing machines, and hundreds of other places.

Education & Technology

A FEW YEARS AGO I attended a conference on using the Internet for education. A speaker caught my attention when he began with a quote: "this new medium", he read, "will be used to educate all of America—no longer will universities be necessary." It sounded great at an internet conference until he pointed out it was said in 1922 or so and referred to the then new medium of radio.

What his quote highlighted is that the internet is undergoing exactly the same transformation as all new media. The first attempt is always to model the new medium on an old one, until the new technology finds its legs. The best example I know of this is a story about William Butler Yeats, the 20TH century's greatest poet.

He wrote to his friends in 1937 calling himself "a fool", "a bore", and —unusual for Yeats—"a humbled man." What defeated this Nobel-prize winning poet? Radio.

From 1931 to 1939 Yeats made eleven experimental radio broadcasts —and learned that radio wasn't a lecture hall. In one of his earliest broadcasts Yeats arranged a celebration of sounds like he performed in drawing rooms and crowded halls across his country. In the broadcast Yeats got a actor to sing his poem and clap his hands in time to the background music after every verse; Yeats also arranged for another poet to lead people in the wings clapping their hands.

The result? In Yeat's own words: "Every human sound turned into the groans, roars, [and] bellow of a wild [beast]." He called it "a fiasco." "It was stirring" in the studio he noted, but "on the wireless [radio] it was a schoolboy knocking with the end of a pen knife or spoon." And he wondered whether "his old bundle of poet's tricks" were now "useless." After that instead of controlling the medium of radio, it began to control Yeats.

When his director told him the opening lines of his now classic poem *Sailing to Byzantium* were easier on the page than the tongue Yeats took his classic open line—"That is no country for old men"— and changed it to "Old men should quit a country."

What Yeats had tried—as did many others—was to use radio as a huge loud speaker, a way to broadcast educational lectures and platform performances. But radio was a private one-on-one communication, even thought the audience was large; it was a medium that turned public speaking into just talking. Radio started by broadcasting long lectures and full-length Broadway plays. But since its audience could move away at the turn of a dial, unlike a theater goer, radio's mix of programming became all important. And soon it offered, instead of hour-long educational lectures, its own a mix of music, talk and soap operas—all tailored to the intimacy and time frame of radio. This is an evolution that's seen in all new media.

Look at the earliest movies and you'll see filmed stage plays— complete with curtain drops because the producers thought audience too stupid to see the relationship between one shot and the next. Only when film came of age did it do things plays could never do.

This is the same transformation we'll somehow see with the Internet. Right now people attempt to make it deliver TV, or newspapers and even radio. But in the end it will be none of these, it'll likely contain bits and pieces of them, but will mix them to form its own style.

Steadicam

I'M FASCINATED BY THE hidden ways that technology affects us. This is best seen—or rather not seen—in the movies.

Here's an example: In *Rocky*—the 1976 Sylvester Stallone blockbuster—the boxer jogs up the steps of the Philadelphia Museum of Art. Only a minute long, this scene made movie directors gasp. What amazed them? The camera work following Rocky up the steps, never once was the shot jerky. It was the work of a then new device, called a Steadicam—the brainchild of Garrett Brown. While the shots from this camera are steady, its development was anything but smooth.

While working at an ad agency Brown noticed how much fun the people filming the commercials had. Brown, who had never even owned a camera, decided on a whim to become a film producer and cameraman. He he bought a thousands dollars worth of equipment from a bankrupt producer and went into business. Brown got the idea for his Steadicam while hanging out of a helicopter filming a car commercial.

They'd hired him to film through the windows of a moving car, but from the outside. Finding it hard to hold his camera steady he resolved to make a special holder. On his way home he stopped at a plumber's shop and bought thirteen dollars worth of pipe. At home he built a crude Steadicam—and from then on he used every spare dollar and minute to refine it. But after two years of work he realized his Steadicam was too complicated.

So he holed up in a motel for a week: No interruptions except for room service. He made a last ditch effort to perfect his Steadicam. He filled notebooks with wacko ideas, occasionally amusing the maids by borrowing their broomsticks to test his ideas. He realized one of the main problems with holding a camera steady is its center of gravity is too far away to control.

You can see this by comparing a suitcase to a backpack. A suitcase is hard to control—its center of gravity is located far from your body, in the suitcase itself, but a backpack is manageable because its center of gravity is near yours. So, Brown developed a belt that anchored the camera to the operator. Next he need to isolate the camera from motion. Brown took the extension arm of his motel desk lamp—the kind that makes a parallelogram—and attached cables and springs to keep the camera from feeling bumps, all the motion being absorbed by the springs.

Movie directors at first avoided Brown's novel Steadicam; they preferred, on a movie set where every minute without filming costs big money, to stick with tried and true techniques. So Brown used his Steadicam to make impossible looking television commercials until he had a demo reel that wowed movie directors. By the end of the nineteen eighties Brown had shot the forest scenes in Star Wars' *Return of the Jedi* and the suspenseful rope-bridge scene in *Indiana Jones and the Temple of Doom*.

For that Brown worked in one hundred and three degree heat, three-hundred fifty feet above a raging river. No wonder he calls using the Steadicam "an artistic, athletic adventure." Noting, it is "a diabolically hard thing to use ... In it's way, the Steadicam" he says "is [like] a violin. Alone, it's just junk; [but] with a good operator, it's magic."

Vacuum Cleaners

THERE IS NOTHING I like more than an engineer with the courage of his or her convictions.

In 1900 the English engineer Cecil Booth startled diners in a London restaurant when suddenly he rose from his upholstered chair, placed a piece of damp fabric on its arm, and began sucking. When done he looked at the resulting black ring of dirt on the fabric with great satisfaction—despite his coughing and choking. Booth was trying to prove to himself that suction—a vacuum—was the best way to remove dirt.

He got the idea at a London train station where he saw a machine used to clean railway carriages. The machine used a huge blower to move the dirt from one end to the other, and hopefully out the door. Booth realized that the opposite process—a vacuum—was needed, and he knew that getting out the dirt was a nineteenth century obsession.

In that time cleaning was an "aggressive art"—housewives used salt, cornmeal, even shredded cabbage to draw out the dirt, then attacked it with a broom or carpet beater. In 1901 he marketed "Booth's Original Vacuum Cleaner Pumps" also known as "Puffing Billies."

They were massive bright red machines. The vacuum power came from an engine so large the unit was moved about on a horse-drawn cart. Booth's team of white-coated operators cleaned carpets, curtains, and upholstery, running hoses from the pump in the street into London stores, hotels, and houses—but only houses of the wealthy because the vacuum was so expensive.

So novel was the work that sometimes the homeowners held tea parties so their friends could see the vacuum cleaner at work. Booth got great publicity for his machine when he cleaned the great blue carpet at Westminster Abbey after Edward VII's coronation. He removed an "immense" amount of dirt impressing the heads of Germany, Russia, and France—all who then wanted a vacuum cleaner. Yet Booth's machine was still too expensive for everyday use. What it needed was a small electric motor, something not in existence in England at the time, but was just appearing in America.

A poor American inventor turned janitor with a hacking cough needed a way to get rid of dust. This man, James Murray Spangler, developed an electrically driven portable machine to suck up dust. In addition to suction Spangler added a rotating brush driven by the motor, and some pillowcases to collect the dirt.

Without enough money to develop the machine he sold it to his cousin's husband William Hoover. Booth and Spangler brought engineering genius to the vacuum cleaner, but Hoover and his son added that essential element to success: sales.

The son became the firm's key salesman. He recalls "I would stock up a hardware store with cleaners, go out two months later and find none of them moved. I would get busy and demonstrate them to housewives and move the stock. Quite unwittingly, I stumbled on the fact that specialty demonstrations were the correct way to sell vacuum cleaners.".

The Hoover's eventually sold thousand nationally with the jingle:

"All the dirt, all the grit
Hoover gets it, every bit."

Although no longer using jingles, Hoover's company still dominates the vacuum cleaner market.

Jet Takeoff

LIKE MANY OF YOU I'm a bit afraid to fly. I'm so alert I listen to every noise a jet makes and, for piece of mind, I've had to learn the source of all these sounds. So, today I'm going to tell you what you hear when a jet takes off. Maybe it will give you some comfort like it does me.

As the jet is pushed back you'll hear the pilot turn on the hydraulic pumps to prepare the breaks for take-off. Then the engines are started with a blast of air from the auxiliary power unit—that is air from the terminal. Inside you'll notice the air-conditioning go off—there is only enough air to run either the air-conditioner or start an engine. As the engines spin they're soon able to suck in enough air on their own to burn fuel and the air conditioning comes back on.

As the jet taxis to the runway the pilots adjusts the flaps—you'll hear the hydraulics again, or if you're near the wing you can see the flaps move. They increase the lift so the jet can take off safely. Once the pilots are cleared for take off you'll heard the pitch of the engines change as the turbines rotate faster, generating more power for take off.

The pilots have worked out three speeds: *Vee one, Vee R,* and *Vee two.* The jet first reaches *Vee one,* the maximum speed at which the pilots can stop the plane. If any error lights go on in the cockpit, or for any reasons the pilots think something is wrong they will bring the plane to a halt. Above about 100 miles per hour the automatic brakes become active. This means that if the pilot turns down the power the breaks kick in. It will be pretty dramatic because the auto-brakes cannot see the end of the runway; they just get the message "stop!" It is the main reason you should keep your seatbelt on, and why you should have a safety seat for your child.

As the jet travels down the runway you'll feel every few seconds a bump if the pilot drives right on the center line: there's a light every seventy-five feet. And sometimes you'll hear the overhead compartments shake, but don't worry they are lightweight plastic and have nothing to do with the structural integrity of the plane. Next the jet reaches *Vee R*, the speed where the planes nose rises. But the jet doesn't take off until it reaches *Vee Two*. At this speed it can take off with one of its engines out. As the jet leaves the ground you'll hear the landing gear come up. In a few minutes you'll hear—and perhaps see —the pilot retract the flaps. They're no longer needed; they are just making the flight bumpy now. Sometimes you'll hear the jet power down slightly—maybe even level off. Likely the pilot is just obeying local noise ordinances and speed restrictions. One thing you won't hear in the first few minutes is the pilot.

The Federal Aviation Administration forbids, for the first ten thousand feet, any talking in the cockpit except that needed for flying. In another few minutes you'll reach a cruising altitude where, as the pilots say, you should sit back and relax. Unless, of course, for you're a fearful flyer like me, and now that you've just taken off you're now thinking of the landing.

Coffee Maker

I TEND TO RAISE EARLY, yet am not really a morning person. My first act is to brew a pot of coffee. As it brews I use the time to contemplate. For years now, every morning, starring at my coffee pot I've contemplated one thing: Why do I have a drip coffee maker, when my parents had only a percolator—a vastly inferior way to make coffee. This started me on a quest for an answer.

After much work I now have the answer ... baseball. I will explain, but first a bit of history about coffee making. Today's machines are automatic drip coffee makers. This is the best way to make coffee, but until now has been too demanding to do well. You start with water that is exactly twelve degrees below boiling, then pour it slowly over ground coffee beans encased in a filter. Usually the filter was cloth, which contaminated the coffee unless cleaned everyday. This was too much for most people, so home coffee making was dominated by the percolator.

The percolator had the virtue of being automatic, but the negative is terrible coffee, often described by coffee lovers as sludge. We all love hearing the "perc" of a percolator, but it's actually a bad thing. The perc comes from the boiling water, which is too hot for making good coffee. Each time the water percs through the grounds the coffee becomes more bitter. A taste so bitter that the percolator drove coffee sales to an all-time low by the mid-nineteen sixties.

In this decade the Bunn Company of Springfield Illinois perfected an automatic drip coffee maker for restaurants. They replaced the cloth filter with a disposable paper one, and they perfected a way to boil the water and cool it slightly before dripping it slowly through the ground beans. This restaurant model made five pots at once, but it opened the door to making a home version.

In 1972 Vincent Marotta of Cleveland Ohio designed the Mr. Coffee machine—the first home drip coffee maker. The key to its success was baseball. Marotta felt he needed a big-name to change generations of percolator users into drip coffee maker users. Marotta who'd played with the St. Louis Cardinals turned to his hero: Joe DiMaggio, the pride of the Yankees.

Now at the time DiMaggio wasn't just an ordinary sports figure—he was a legend with a bit of mystery due to his brief marriage with Marilyn Monroe. Marotta somehow got DiMaggio's unlisted phone number in San Francisco and called one Saturday morning. DiMaggio answered. After Marotta pitched his ad campaign DiMaggio told him he'd just won a Mr. Coffee machine in a golf tournament and said his sister ``is making coffee with it right now." Yet the ex-baseball star wasn't interested in being in a commercial, saying ``I don't do that kind of work."

The next day, Marotta and his wife flew to San Francisco, where Marotta called DiMaggio again and invited him to lunch. DiMaggio agreed. During lunch DiMaggio's expression barely changed when they shook hands after the meal, he agreed to make the commercials. And those commercials, with the integrity of Joe DiMaggio behind them, put Mr. Coffee on the map—and coffee makers into nearly every American home.

Grain Elevators

WHEN MY WIFE and I moved from the East to the Midwest we found the landscape dull. We missed the beauty of rolling hills, mountains and rivers—and as an engineer I missed tall buildings and complex bridges.

After a few months my wife looked at me and said "the beauty is here, the trick is to look carefully." And indeed she'd found beauty in the bloom of wild flowers and in glorious sunsets. Sunsets that dazzle because the flatness of the plains lets the whole sky be a canvas. Still I found few buildings that fascinated me, until I noticed grain elevators.

You've seen them: They look like a chimney standing alone. In the early twentieth century Le Corbusier, the great Swiss architect, called them "the magnificent first fruits of the new age." He felt they defined the times, along with planes, cars and ocean liners. Now, of course, planes and cars are symbols of decadence, and ocean liners of obsolescence, but grain elevators are still with us, their reputations intact.

Le Corbusier admired their unadorned purely functional concrete exteriors. Yet, so silly was the idea of using concrete to build a grain elevator that the first, built in 1900, was known as Peavy's Folly.

Frank H. Peavy owned a granary in Minnesota just when the Midwest moved from subsistence farming to cash crops, which caused an explosive growth in the number of crops shipped by rail. Peavy at first transported his grain in sacks, but this was expensive and time consuming. He found it quicker and cheaper to load his grain by dumping it into the railroad cars. Hence the name elevator—to elevate the grain above the railroad. Peavy tried building wood elevators, but sparks from the engine frequently light them on fire. He was impressed by concrete bridges, so he hooked up with Charles Haglin, a local builder, to make a concrete grain elevator.

The two men faced opposition because engineers across America argued that a tank of solid concrete would not have any "give" and would thus explode when the grain was taken out. Undaunted they moved forward, building their grain elevator like a cake: First pouring a ring of concrete, letting it dry, then adding another ring on top. Eventually they built a hollow structure sixty eight feet tall, and twenty feet in diameter with walls about ten inches thick. When it dried they shoveled grain in and let it sit for the winter. In the spring as they prepared to let the grain flow out a huge crowd gathered, although they stood back a good block from the elevator sure that it might explode.

With the courage of his convictions Charles Haglin stood at the elevator's base and pulled the lever to release the grain. It was in perfect condition. And the tower didn't explode. It still stands near Minneapolis. This first circular reinforced-concrete grain elevator pioneered the way for the thousands across American's heartland.

So important is this particular grain elevator that in 1982 it was placed on the National Register of Historic Places.

Jekyll & Hyde

THE WORDS *engineering* and *literature* aren't often used together, yet an engineer's life is the source for Robert Louis Stevenson's *The Strange Case of Dr. Jekyll and Mr. Hyde.*

The story came from Stevenson's fascination with duality, of how good and evil coexist in the same person. Dr. Jekyll is a large, benevolent physician, although not entirely a good man. Chief among his human frailties is his foolhardiness which causes him to release from inside himself a small evil man, Mr. Hyde. For Stevenson the fascination was not that Jekyll changed into this evil man, but that it was already inside him.

At one point Stevenson has Dr. Jekyll look in the mirror and say of Mr. Hyde "And yet when I looked upon this ugly idol in the glass, I was conscious of no repugnance, rather of a leap of welcome. This, too, was, myself." Stevenson's inspiration for this was the life of an engineer friend. Stevenson wrote Jekyll and Hyde while mourning the sudden loss of his much loved friend and mentor Fleeming Jenkins, an engineer. Jenkin's sudden death shocked Stevenson and to help him grieve he began a biography of his friend, but shortly after beginning he stopped and took ten weeks to write *The Strange Case of Dr. Jekyll and Mr. Hyde.*

It's obvious that as he wrote engineering was much on his mind. At the surface Jekyll and Hyde is filled with allusions to prominent engineers.; his alter ego was named after Major General H. Hyde, a Royal Engineer and member of the Society of Telegraph Engineers. The evil Hyde's victim shares a name with another Telegraph engineer, whose murder in the story is investigated by Inspector Newcomen, the same name as the famous inventor of the steam engine.

This naming game is playing throughout the novel, yet there is a deeper connection between engineering and Dr. Jekyll and Mr. Hyde. Stevenson shared with his friend Jenkins a love of drama and poetry. Yet never was their bond through engineering. Stevenson wrote that Jenkins "taste for machinery was one that I could never share with him, and he had a certain bitter pity for my weakness." Stevenson said that to Jenkins the "struggle of the engineer against brute forces and with inert allies was nobly poetic." To Stevenson his engineering friend Jenkins was an embodiment of two contradictory things— engineering and poetry. In this Stevenson saw a duality; he saw a divided self, a double life.

How, he wondered, could a love of engineering and poetry exist together in the same person. So, as Stevenson wrote a biography of his late friend Fleming Jenkins he probed the duality of his engineering friend. And from this exploration of how two people could be contained in a single person emerged his greatest imaginative work *The Strange Case of Dr. Jekyll and Mr. Hyde.*

Lava Lamp

A S A BABY BOOMER I'm amused to see the era of my youth—the 1960s and 70s—returning in TV sitcoms and movies. As a survivor of those decades—and as an engineer—I'm going to jump onto this 70s band wagon and share with you the details behind an icon of the era. In engineering terms I'm talking about thermal chemical convective motion of strongly temperature and depth dependent viscosity liquids. No, it isn't a story about drugs, it's the Lava Lamp.

I recall as a kid seeing Lava Lamps on the *Avengers*, the British TV series. The Lava Lamp was the brainchild of Edward Craven Walker a former Royal Air Force Squadron Leader in World War II. After the war he started a travel agency—and he indulged in a lifelong passion in what the British call "the Natural Way of Life." Edward Walker was a nudist.

While on a "natural" vacation off the coast of France he made films promoting this way of life. A film of the naturalists doing underwater ballet to Rimsky-Korsakov's *Song of India* made him a fortune which he used to promote nudism—and to develop the Lava Lamp.

While visiting a London pub he noticed a heated glass cocktail shaker filled with oil and water. The rising and falling oil entranced Walker so much he immediately bought the rights to produce this lamp. A Lava Lamp seems a simple thing, but to make one work correctly isn't a simple thing at all.

The Lava Lamp is filled with globes of oil in water so that as heat is applied to the bottom of the Lamp the heavier liquid becomes less dense and rises; as it rises it cools becoming more dense and falls again. Getting the proper mix of liquid is the main secret of Lava Lamps. Edward Walker worked in his back yard for ten years to perfect the proper formula—its a mix of water, oil, wax and other solids that is still secret today.

Walker's Lava Lamp caught on quickly—even though the large British Department store Harrod's said it was disgusting and refused to carry it. Walker sold it with the slogan "If you buy my lamp, you won't need drugs." To Walker the Lava Lamp's motion was "like the cycle of life. It grows," he said, "breaks up, falls down and then starts all over again." "Besides," he added, "the shapes are sexy." This, of course, passed for philosophy in the 1960s. His lamp took off as psychedelia took hold in the mid-1960s. The Lava Lamp is perhaps the symbol that most characterizes that era. So much so that at a recent ceremony by the 70s Preservation Society—a group that keeps that decade alive—they skipped the usual statutes for their annual achievement awards and just gave the winners Lava Lamps.

Pumpkin masters

SOME PEOPLE TAKE THEIR work home with them creating tensions in their families, but one man Paul Bardeen brought his work home and changed Halloween for his family and for millions of Americans. Bardeen, an electrical engineer, worked at the state power company in Wisconsin. As their chief safety engineer he developed ways to prevent accidents. It is this safety-mindedness that Paul Bardeen brought home to his family. Like most parents, maybe even more so, he did all the usual safety things: He made sure all the hand rails at home were sturdy, zealously kept the steps clear of ice, and lectured his children about driving defensively.

But the difference between Bardeen and other parents was his approach to Halloween. In spite of his careful demeanor Bardeen was a gregarious man who loved nothing more than a pumpkin carving party with his five children. Yet here was a problem for this safety engineer: Five kids each with a knife stabbing away at their pumpkins, risking puncture wounds and lacerations.

Bardeen also noticed that the knives didn't let his children be creative enough; a knife was too blunt an instrument to carve anything interesting. Bardeen the safety engineer retired to his workshop to devise a safer, yet more creative way to carve pumpkins. He returned with an intricate drawing of a Jack-o-Lantern face. He showed his children how to attach this drawing to a pumpkin and then take a nail and press it though small holes he'd make in the pattern to transfer the image to the pumpkin. Then he gave each child a special saw he'd made in his shop: A coping saw blade with a dowel for a handle, sized just right for a child's hand.

This method allowed Paul Bardeen and his family to create the most fantastic of Jack-o-Lanterns: One with witch's brooms for eyebrows, a bat for a mouth, and ears shaped like cats. When trick or treat time came Paul Bardeen's children carried their pumpkin masterpieces door-to-door to impress and amuse neighbors. Paul Bardeen died in 1983 and in his honor his son John started a company called Pumpkin Masters that sells the tools and patterns developed by his father to help children more safely carve Jack-O-Lanterns.

After a rocky start John Bardeen's Pumpkin Masters now sells millions of kits across American bringing his father's unique blend of pumpkin carving and safety engineering to all children.

Video games

THE LATEST IN VIDEO GAMES, the Sony *PlayStation2* is out, and, as always stories abound about long lines and scuffles. Some people have even camped out overnight to buy the first games sold. These night-long waits pale compared to a man who waited twenty years for his piece of every new video game. He's Ralph Baer, the inventor of the idea of the video game. His video game story begins in the 1940s.

Baer, a refugee from the Nazis, came to America and earned a degree in television engineering. In 1951 Baer was given the task of designing a new television set. Although he toiled for months the set never went into production, but Baer learned all about how TVs work.

Soon he moved out of the television industry and worked for a military contractor making all sorts of electronic gadgets that kept our armies at the ready. While waiting at a New York City bus terminal Baer, bored, let his mind rove.

Using what he'd learned while building a television set he conceived of a way to control the images on a TV screen so he could play games on it. When he got back to his office he wrote a four page memo listing the types of games that could be played: action games, board games, sports games, chase games, and so on. Under the guise of inventing a training tool for the military he designed a circuit to control two spots on a TV screen.

With help from a colleague, he built it, hooked it to a TV and then he and his co-worker played a "Chase" game: One spot was a fox, the other a hunter. Baer lost. Over the next year he refined this game and pitched it to television manufacturers.

Magnavox bought the idea and by 1972 brought out the first commercial video game. Called Odyssey it offered just about any combination of things you could do with a few blips: volleyball, handball, and wipe-out.

While modestly successful, video games became immensely popular when Magnavox's competitor Atari brought out Pong. But Ralph Baer and his company held the patents on the idea of a video game. Baer spent the next twenty years in Federal Courts in Chicago, New York, and San Francisco. He won every case. So, in the 1970s and 80s Baer and his partners sold licenses to the big producers: Atari, Nintendo and Sega.

I sometimes wonder what Ralph Baer thinks of the latest Sony *Playstation*. I thought about giving him a call, but likely I'd have to pay him. Today, at age eighty—some thirty years after he invented the video game—Ralph Baer still runs his own consulting service telling manufacturers how to make better video games.

Typewriters

I GOT A CATALOG IN THE mail yesterday filled with the latest electronic gizmos. One ad caught my eye. The headline read: "A machine that performs tricks a computer can't."

Intrigued I read on: "Simply plug it in, switch it on, and its instantly ready," which the ad noted is "amazing" when compared to how slow a computer starts. What is this amazing machine? Its a typewriter. I love this because I still own—and use—three manual typewriters. Still I know its nearly time to write their obituaries, but before I do I want to give the typewriter its due.

The earliest proposals for typewriters are from the 14TH century, but its influence really began in the 19TH, increasing in the 20TH century until it created a revolution. From the 14TH century to the mid-19TH some 51 people tried to build a useful typewriter, but the key engineering insight came from the 52ND person to invent the typewriter: Christopher Sholes in the 1860s.

He realized that the paper, not the letters should move. Sholes hooked up with the Remington Gun Company to make typewriters. This sounds odd, but after the Civil War their sales were slack, and they'd already moved into sewing machines, so why not typewriters. The original market Sholes and Remington aimed for was ministers and writers—hopping it would later catch on with the public—but they did not, as would seem obvious, aim for businesses. They backed into the business arena when they began marketing, in desperation, the typewriter to the daughters of middle-class businessmen.

The ads proclaimed "No invention has opened for women so broad and easy an avenue to profitable and suitable employment as the typewriter." The typewriter truly opened the world of business to women. Five years after the YMCA introduced typing classes there were 60,000 female typists in the U.S. Not, of course without an outcry: Women in offices would make the family collapse. And also women were too weak for the manual labor of typing. The YMCA countered this by requiring "a thorough physical examination of all the applicants."

The typewriter also revolutionized other aspects of our 20TH century. My favorite is how the typewriter changed poetry. Prior to the typewriter poetry was written for the ear only, but with the typewriter a poet could write for the eye, laying out his or her words in a precise way—expanding meaning to include white spaces, not just words. This led the way, for example, to the wonderful poems of e.e. cummings.

There now I've given the typewriter it's due. Let the day and age of the computer continue, but let's see if it brings us another revolution in the work force—or a poet with the power of e.e. cummings.

On-line shopping

YESTERDAY MY WIFE AND I helped break down the social order. We're not revolutionaries, but just two placid engineers. What we did was this: We bought groceries.

Not at the store, but via the internet. Our local store has a web site where we just click and then the next day the groceries appear on our doorstep. Now I say "break down the social order" because ordering via the internet takes us out of contact with people. It doesn't foster interactions and build communities and relationships. I came to this view while touring Amish farms with my uncle.

He lives in Indiana, and as a realtor has sold farms to Amish settlers from Pennsylvania. As we drove past the new farms he pointed at one and said "the Amish farmer who lived there publishes a newsletter, he writes it on his laptop." Laptop? Amish? Yes, I learned. I thought the Amish were anti-technology, but I learned that their approach is not disdain, but wariness. They watch how a piece of technology affects others in the "outside" world and then with caution bring it into their lives. For example, the laptop computer can be used only in the barn —it isn't to invade their home. Now I'm not saying the life style and methods of the Amish would solve our ills.

In fact, clearly not everyone in our world of five billion could live this way, but there is a message here: carefully analyze any technology before adopting it. In my field there is a saying: "Technology is neither good nor bad; nor is it neutral", meaning that each aspect that touches our lives must be examined.

I took time last week to observe this adage in practice. I focused on how technology insulated me from people. For example, I noticed I got cash from an ATM, instead of a real bank teller. Or, I used e-mail extensively to talk to my co-workers. There were few face-to-face meetings. And when I called information for a telephone number I got a recording. The phone company equipment taped my request, processed the sounds and fulfilled my request by computer.

Now I've been very negative about using electronic communications, but I'm certainly not anti-technology—my wedding ring proves it. The most important thing to me about this ring, this most traditional of symbols, was that if I lost it I wanted it returned. So I had my jeweler engrave not my wife's name, or the date of my marriage, but instead he engraved my e-mail address.

Project Gutenberg

L AST NIGHT I WANDERED my home in search of a novel by
Henry James. I keep thousands of books in my home because
often late at night I'll get the urge to read a particular book or author.
But my shelves held no books by James. Long ago I'd declared them
dull and tossed them out. But now, at midnight, I needed to try again.
How to satisfy this urge? I turned to the internet and browsed *Project
Gutenberg*.

This on-line archive of thousands of books had, for free, all of the
Henry James my heart desired. I learned just yesterday that *Project
Gutenberg* began in my home town. Michael Hart began it at the
University of Illinois in 1971.

One day Hart found his computer account credited with one
hundred million dollars worth of computer time. Now, most of us
would just report this error, but Hart thought differently. He wanted
to repay that gift. An idea struck him later that day as he shopped for
groceries.

As they bagged his groceries they tossed in a copy of the *Declaration
of Independence*; it was part of the growing mania for American's
bicentennial. Hart realized that if he typed this into his computer and
sent it to everyone, copies of it would exists everywhere and forever.
Hart reasoned that he'd have repaid his one hundred million credit
when there existed one hundred million computers all storing his
typed in *Declaration of Independence*.

So, he set down and began typing "In the course of human
events ...." after finishing the last word he sent it to everyone with e-
mail—now in 1971 this was only a couple of hundred, but still this first
piece of junk mail upset everyone. To avoid this in the future, Hart
just posted it, that is put it on a computer where others could retrieve

it when they liked. What Hart realized was that the major impact of computers wasn't in calculating, but in distributing. This started Hart on his quest to build an electronic library. He typed in the *Bill of Rights*, the *u.s. Constitution*, Shakespeare's plays—setting for himself a goal of ten thousand books on-line by 2001.

None of this work was funded; Hart did it all himself until 1988 when the internet reached a quarter million users and his *Project Gutenberg* attracted others. Today about one thousand volunteers type, scan, and proofread about 150 new books a month.

There are now some three thousand books on-line; a bit short of his goal of ten thousand, but note that some 100 million people can now access his books. And perhaps a trillion copies now sit on computers all around the world. This is Michael Hart's gift to the world.

So from me, at least, thank you Mr. Hart for the gift of Henry James at midnight.

eBooks

I'VE STARTED TO SEE ADS for electronic books, called e-books. These ads insist that paper based books are out of date. They'll be replaced by an e-book which looks like a small laptop computer. The ads promise a "new way to enjoy reading." Now as a dedicated reader —I have some two thousand books in my home - I wasn't aware the old way had any problems.

As an engineer I'm going to make a rash predication about technology: E-books will flounder at best.

It's mostly because they won't appeal to core book buyers. By core I mean the seven percent of the population that buys fifty books or more a year, compared to nearly half of the public that buys fewer than five books. The main problem with e-books is they focus only on content. The ads promise you can "read comfortably on a large, clear screen", but they miss the tactile aspects. I didn't fully realize this facet of reading until my wife pointed out that after I turn a page I slowly run my hand down the center of the book - caressing it in a sense.

Touching and holding the book helps to communicates its information: I can remember if a phrase was on the left or the right hand page - and my hands tacitly tell me just how far into the book the information was located. All of this lost with an e-book where each page is tactilely identical to the previous one.

Next, the ads promise "thousands of titles." But getting thousands of titles isn't ever my problem, it's getting *through* thousands of titles. The e-book ads proclaim that when you travel you can have "dozens of books" at your fingertips. But I've found over the years that the secret to reading a lot is to take only one book with me. If I take two I never get much read; I spent the time choosing which book to read, but with one book I have no choice and so I plow through it on the trip.

I'm not upbeat about e-books, but I do think technology is going to revolutionize the publishing industry. Here is where I'd place my bet. Usually technology impacts us the most when it blends the new and the old together. So, keep your eye on the machines, which are just beginning to appear, that print books on demand. It's a very fancy laser printer with a binder attached. Choose a title, press a button and out pops a paperback not much different than one you might buy today. This radically changes the distribution and cost of books, and also leaves the book lover with exactly what he or she wants: A real book to touch and hold.

Scotch tape dispenser

YESTERDAY, I BOUGHT the essential tool for getting through the holidays: Scotch tape. As I purchased it I marveled at its plastic, circular dispenser, the one that looks a bit like a sea shell. The design's durability amazes me - it's sixty years old - and it's what made scotch tape really happen.

Richard Drew, a chemist working in the 1920s at 3M, invented Scotch tape when a company planning to insulate railroad cars approached him. They needed to protect their insulation from water because it stunk when wet. A friend suggested Drew try a new material called cellophane.

So he took cellophane, coated it with adhesive and, although, he didn't solve the insulators problem, he invented scotch tape.

The tape's remarkable ability to stick to everything attracted all types of users, but it stuck too well to itself: Users couldn't keep the loose end of the tape from reattaching to the roll, where it became invisible and nearly impossible to get a hold of again. To really take off Scotch tape needed a proper dispenser.

A tape dispenser seems trivial, yet it requires a deep knowledge of psychology and physiology, of science and technology and an understanding of manufacturing. Here enters the least known of all people who effect our material world: The industrial designer.

In 1937 3M hired designer Jean Reinecke to create a new holder. Reinecke spent the next forty years thinking about scotch tape dispensers. He made two designs before hitting in 1939 on the first of his seemingly permanent contributions to our material world.

His 1939 dispenser - the classic I bought recently - is the simplest holder: Two plastic pieces that press together, the roll held in a circular section from which an arm extends to hold a serrated edge to cut the tape. To me the design is so simple it looks fresh even today: sleek and elegant in its geometric precision. It fits perfectly in the palm, keeps the tape at the ready, yet can be easily and cheaply manufactured.

Despite the success of this holder Reinecke kept thinking about tape dispensers: In 1953 he designed the classic cast iron office tape dispenser: An elongated blocky base with circular shrouds, gently sloping sides and a raised blade like a violin bridge. So durable were these that many originals are still in use today.

Like the best designers Reinecke kept up with the times: For the 1960s he designed a highly sculptured, molded plastic form with swooping curves, available in hundreds of day-glow colors. I still have one in my office. His tape dispenser designs are now permanent icons of mid twentieth century design - and are likely to stick with us through the 21st century.

Velcro

IN THIS WINTER SEASON HERE in the states, I have renewed appreciation for my wife's greatest talent: Choosing coats. When my wife picks out a coat everything from style to utility is perfect. The material is ideal for keeping the wind out, every fastener is perfect: a drawstring for the waist and neck, a zipper and snaps for the front, and for the cuffs, Velcro. I'd argue that without Velcro my wife couldn't buy a perfect coat.

Velcro's the invention of a Swiss engineer, George de Mestral who came up with the idea while walking with his dog in the woods. de Mestral observed that his wool socks and his dog's fur were covered with burs. When he got home he examined the burs under a microscope and saw that their barbed, hook like seed pods meshed with the looped fibers in his clothes. Here, he thought, is an ideal fastener. His idea met with resistance and even laughter, but de Mestral, with help from a weaver and a loom-maker, perfected his "hook and loop fastener." He worked to make a synthetic material that duplicated the burs clinging to his wool socks.

By trial and error, de Mestral realized that nylon, when sewn under infrared light, formed tough hooks for the bur side of the fastener. The difficulty of attaching hundreds of tiny hooks to cloth tape held up de Mestral's work for eight years, and then mechanizing the production process of weaving 300 hooks and loops per square inch was another hurdle. Finally, by 1955 his "hook and loop fastener" was patented under the name Velcro, which he formed from two French words *velour* and *crochet* - velvet and hooks. The shear number of places that Velcro now appears is astonishing.

In medicine: The two pumping chambers of the *Jarvik-7* artificial heart were fastened with Velcro, and how were blood pressure cuffs held together before Velcro?

The success of every Space Shuttle flight depends on it: Everything in the interior - food packets, tools, even astronauts at times - are held down by Velcro.

And, of course, it's held up David Letterman, the late night talk show host. He donned a pair of coveralls made of Velcro hooks, bounced from a trampoline into a wall covered with Velcro loops, where he stuck. Now that the Velcro company has covered everything from outer space to talk show hosts, what's next?

Maybe our stomachs. They've discussed edible Velcro: Think how useful this would be to keep closed a burrito or the pancakes of moo shoo pork.

Spam

UPON MARRIAGE I RECEIVED my wife's dowry: a can of Spam. The blue, rectangular tin has been with her for fifteen years. A friend gave it to her as a reminder of a Monty Python skit:

> *"What you got? Spam, eggs, Spam, Spam, bacon and Spam. Have you got anything without Spam? Well, the Spam, eggs, sausage and Spam. That hasn't got much Spam in it. I don't want any Spam."*

Spam was the brainchild, in 1937, of Jay Hormel. It began ten years earlier with an idea crazy for its time: Canned Ham.

In 1926 all meat products perished quickly, this suggested to Jay Hormel that he could make a fortune if he packaged meat to last forever. For three years Jay Hormel and his workers tried to make these crazy canned hams, but they couldn't seal the tin cans well enough. As his workers struggled Jay Hormel went abroad for the summer. In Germany he learned of Paul Joern, whose meat packing business had gone belly up, but who owed international patents on canning meat. Hormel asked Joern to come to Minnesota; he jumped at the chance to leave his business troubles.

Joern showed Hormel a secret way to solder the can shut to make what became "Hormel flavor-sealed ham." Where does Spam come into all this?

Well, Jay Hormel's crazy canned hams sold well - mostly because of a half million dollar advertising campaign - but within ten years every meat processor made canned hams, and Hormel began loosing market share. He realized that he needed a brand name he could copyright. He also noticed that his company wasted a great deal of meat by tossing out the shoulders of the hogs. So Jay Hormel devised a product using the shoulder - in fact the "mystery" of what is spam isn't very interesting, its meat from the shoulder, with some ham, salt, sugar, nitrates and water. Hormel couldn't call this "ham" because the government restricted "ham" to be a hog's rear-quarters.

To name his new product he held a party where he gave guests a drink for each name they offered. He recalls that "along about the fourth or fifth drink they began showing some imagination" and someone tossed out "spam." With that name Jay Hormel began an immense advertising campaign, even imploring his stockholders to have a "high style" ... "breakfast of Spam and eggs, [or] eat a Spamwich at noon ...." Today the word "spam" brings forth laughter, but it was - and is - a real success story: Five billion cans of Spam have been eaten since 1936.

To celebrate, the Hormel Company has built a Spam museum in Minnesota where you can see Spam through the ages, or eat Spam in the cafeteria, or buy a Spam hat, t-shirt, watch, golf ball, glassware, water bottle, mug ....

Ultrasound imaging

I FOUND, THIS MORNING, in my e-mail in box an amazing image: A friend had sent an ultrasonic image of her developing baby. Today we usually associate ultrasound with the joyous event of birth, but this magnificent technology came about because of the horrors of World War II - and because of a London surgeon named John Wild.

He operated on civilians with internal injuries from German Vee one rockets. After hundreds of operation Wild realized that to increase the survival rate he needed to know how the intestines were healing; he needed to see whether the wall thickness was increasing or decreasing. The only way he could do it was to operate again, which was very risky.

After the war he moved to Minnesota and devoted his full attention to figuring out how to see inside the human body. He'd heard of pulse-echo equipment used in the war to detect cracks in armor plate - it worked like a bat's sonar - but he couldn't adapt this equipment because the cracks were much smaller than an intestine wall. He asked the equipment's maker to estimate the cost of a custom unit with the appropriate resolution for medical use. Their answer: one hundred thousand dollars - a very steep price in the late 1940s - and way too much for just testing the idea. But John Wild persevered. He searched until he found a physicists who'd developed an ultrasound device used to train airborne navigators to recognize radar images.

The machine "flew" over a small relief map detecting the tiny ups and downs of hills and then reproduced them on a radar screen. These ups and downs on the map were just the right size for imaging inside a human body. After the war this piece of equipment lay idol at a nearby Naval Base in Minneapolis. Wild talked its owners into letting him use it.

His goal was just to test the idea, so he brought over a dog's small intestine to image. The result was good enough that Wild could move forward with building his own equipment. Unable to test his new ultrasound machine on a person he instead peered inside a pie. His wife had some fresh steak and kidney pies in the frig, so Wild borrowed a slice and examined the inside of the pie without cutting it open. As he played with his ultrasound machine he realized by 1949 that he could use it to detect cancer - a completely novel idea at the time.

So, by 1951 he used his ultrasonic technique to examine real patients, focusing on detecting colon and breast cancer. Today John Wild's imaging is used in thousands of ways, to save hundreds of thousands of lives. And its promise isn't yet completely fulfilled. Researchers are now starting to study ultrasonic images of the flow of blood in our brains. Their goal: To understand how we think.

Gas Lighting

NEARLY EVERY MORNING I RISE AND sit by my flickering fireplace reading for half an hour or so. Recently, I learned that my neighbors were impressed because day after day they saw, by seven a.m., that I'd built a fire. I finally confessed that it was just a very realistic gas log; my only industry was to stumble out of bed and turn a knob to ignite the gas. This same gas-powered glow also impressed neighbors some two hundred years ago. In Newport, Rhode Island a house on the corner of Pelhman and Thames streets gave off what was then considered an eerie glow - too strong to be candle light, and too ubiquitous to be an open fire in every room.

It was the house of David Melville. By day he was a hardware merchant, but by night he was an inventor who experimented with lighting his house using "inflammable air" - what we'd call gas. He'd built his own gas generating plant in his basement: A small furnace that burned coal and made his so-called inflammable air. He stored the gas in a tank and from there he piped it around his house to be burned in special fixtures.

In 1813 he announced a public demonstration of his gas lighting with the intent to sell his system to every American house. The demonstration's purpose, he said, was to "gratify public curiosity" and allow him "to be to some degree remunerated for the very great expense" of his experiments - so he charged a fee of twenty five cents. He assured the curious public that gas lighting was "in no way offensive" and that it was safe. Melville's demonstration brought people by the thousands, followed by huge orders from factories in Providence and Watertown, Massachusetts. But once they learned the cost of installation and operation these customers canceled orders. With no prospects for home or industrial lighting Melville turned his attention to lighthouses. He got a federal contract to develop gas powered lighthouses, but the government withdrew it under pressure from whale oil interests - the principal suppliers of oil for lamps.

So, Melville went back to his pewter trade and hardware store, but he'd lit a fire that could not be quenched. Within a few years gas lighting burst upon the scene with profound social and economic consequences. It lengthened the working day and made streets safer at night. It made it easier for dinner, usually taken at three in the afternoon, to slip into the evening. And evening classes became a possibility, allowing working people to gain an education after the day's work was done.

And now it makes it possible for me to turn a knob and have a roaring fire without chopping any logs.

New vs. Old Technology

T HERE IS AN ILLUSION ABOUT TECHNOLOGY that whatever is the newest and most different succeeds. Yet to really catch on a technology must blend the old and the new together. You can see this best in automobiles.

My father bought a new Oldsmobile every five years or so - and when he hopped into his new one it felt like his old car, with all the controls in the right places. The main engineering principle is that to succeed you must combine the old and the new. I learned this principle indirectly from my father years ago.

He sent my sister, brother and me weekly envelopes filled with clippings, articles and all sorts of odd things. He reached his pinnacle when I was a college freshman. An envelope arrived and I opened it expecting the usual clipping but found a sack. A brown paper lunch sack folded neatly with a scrap of paper clipped to it: "Here," my father typed, "have a sack." There was no hidden message here, no joke he just thought I might need a lunch sack. The note was his key signature.

He'd grab any odd-shaped scrap toss it into his trusty Royal Typewriter and tap out a masterpiece of condensed prose: "Weather fine, enjoying back porch, here's an article. Love, Dad." About a year after the sack - in 1981 - my siblings and I bought Dad a computer.

We returned to our dorms anticipating his letters. Yet, soon the same typewritten notes arrived. "Use your word processor" we implored. Soon even his typewritten notes trickled to zero, and never did a word processed note arrive. Finally, I asked "Why don't you use your computer?" "Because," he responded, "when I turn it off, what I'm writing vanishes and I've got to begin all over again." He had no concept of save. Here was the solution: Explain save. But then I thought of how happy he was to send out these short notes written on a scrap, and how happy we were to get them. I looked at his printer seeing how we'd locked him into a white 8 ½ x 11 sheet - in those days the printers could use only a roll of perforated sheets. Today, nearly any sized scrap can be run through a printer, but in 1981 printer technology zapped my father's creativity - it took away his medium. He said to me again "my letters vanish when I turn off the computer." I bit my tongue and said "yeah, isn't that terrible. You might as well use your typewriter." He never again turned on his computer.

And to the end of his days he filled our mailboxes with odd-shaped missives of love: clippings, notes on scraps, and even a sack or two. And I've kept that Royal Typewriter to remind me of this engineering lesson.

Glass

TODAY WE MARVEL AT HOW high tech gadgets change our lives, yet right under our noses is an innovation that truly changed our world, yet is so mundane we rarely even see it: It's glass. Of all technological achievements glass changed our world the most.

In the Western world, it first appeared in the 13TH century, but its high cost allowed glass to transform, at first, only public buildings.

How spectacular it must have been for a 13TH century parishioner to enter a church and see for the first time a sanctuary lit brilliantly by light filtering through stained glass windows. Glass gave churches even more magnificence than the carvings and gold that filled windowless baroque churches. As the price of glass dropped it spread from churches to homes.

Imagine how it changed life for a homeowner in the 16TH century. Before glass, windows were sealed with wood shutters or oil cloth and muslin. They isolated the homeowner from the world, the glass windows allowed the sun in, even on a rainy or cold day.

So precious at first to the homeowner was the glass that before leaving home the panes were removed and placed in a safe place. Soon glass begin changing in more profound ways how we viewed the world. Lens for spectacles first appeared as bifocals. Some scholars credit this innovation with an increase in learning because of additional years of eyesight for reading.

Then glass revolutionized further our view of the external world by extending our field of vision. At the turn of the 17TH century the telescope and the microscope appeared - devices loaded, of course, with glass optics. They extended our view from the vanishing point to the edge of the cosmos.

Next glass brought about the modern world we live in. Our lives today are revolutionized by materials: We are surrounded by plastics, polymers, composites and all sorts of chemicals. Glass made this possible: without it chemical reactions could not be studied and analyzed. Glass is an ideal container: it's resistant to most chemicals, neutral in any experiment, yet lets a chemist observe transformations - a combination no wood, metal or clay container can rival.

From this glass made possible other life changing inventions: the barometer, the thermometer, the electric light, and the first electronic devices, filled, of course, with glass vacuum tubes.

So ends my ode to glass: It keeps us isolated from the elements, but lets the sun in; it's given us sharper eyesight, and allows us to take a visual journey from an amoeba to the stars - and lets us recombine the chemical elements in a billion different ways. Now isn't that more marvelous than the latest electronic gizmo?

Super Soaker

I KNOW IT'S A CLICHÉ TO SAY of an invention that it takes a rocket scientist, but it did take one to bring the squirt gun into the high tech world.

I'm speaking of the Super Soaker - the most popular toy of the 1990s. For those who haven't seen it, it's been described as "a squirt gun on steroids." It holds some two gallons of water, and can drench an opponent up to forty feet away - enough to send anyone crying back to mommy. It was invented by engineer Lonnie Johnson, who worked at NASA's Jet Propulsion Lab. He'd designed, for example, the power supplies for the Galileo space probe.

One day he was experimenting with a new type of refrigerator that used water instead of freon. He hooked a nozzle to a faucet in his bathroom - as he turned on the water it shot across the bathroom, making air currents so strong his shower curtain started to swirl. His first thought, and I quote him here, was "Boy, this would really make the neatest water gun." His key squirt gun insight was to use pressurized air to drive the water through a narrow hole in the nozzle.

"From that point", Johnson said building a high tech squirt gun "was an engineering problem." Where the engineering came in was to come up with a way that "a small kid would be able to pump the gun up to a very high pressure." Johnson went to work in his home workshop.

Using a small hobbyist's lathe, he built a model out of PVC pipe, an empty plastic Coke bottle and Plexiglas. Next up was test marketing: He let his daughter, age six, try it out on neighbors. The result: A great success, at least for her. Next Johnson had to interest a manufacturer.

He first approached Daisy Manufacturing, the maker of BB guns, but they passed on his idea after two years of discussion. In 1989 Johnson met with the Larami Corporation. He walked into the meeting, opened his suitcase, and pulled out his prototype of PVC tubing, Plexiglas and plastic soda bottles. A split second later, he fired a giant stream of water across the room. Larami's president had just one word: "Wow!" But would they sell? In the past a squirt gun sold for twenty-nine cents: Would any one pay ten dollars for a squirt gun?

The first year startled the industry. Sales took off when Johnny Carson on the *Tonight Show* used a Super Soaker to drench Ed McMahon. A year later, it was the most popular water gun in American retail history - sold not only by toy stores, but by upscale adult stores like Sharper Image. By the late 1990s about 250 million Super Soakers have been sold -- enough for each person in the United States.

Contact Lens

DOROTHY PARKER ONCE WROTE - in a now famous couplet - "Men seldom make passes / At girls who wear glasses." Some credit her with bringing about contact lenses, but I give the nod to a Czech chemistry professor named Otto Wichterle.

In 1952, while on a train ride to Prague - Wichterle was a citizen of Czechoslovakia - he observed a fellow passenger reading about metal implants for eyeball replacement. He told the traveler "It would be much better to invent some plastic for implants that would be compatible with the surrounding tissue."

Wichterle was a chemist who studied polymers - long flexible pieces of plastic - and he was fascinated with making polymers compatible with the human body. His fellow passenger turned out to be the secretary of a health ministry commission looking into the use of plastics for medicine. The commission ordered Wichterle to make such an implant, but embarrassingly he had to tell them that he had no such material - adding quickly that he was sure he could synthesize some.

He started looking at polymers called hydrophilic, or water-loving - and even before he started, he began filing out patents on potential uses, one of which was for soft contact lenses. At that point his life changed dramatically: The Czech Communists party dismissed him from his job calling him politically unreliable.

Wichterle focused his newly found free time on finding ways to turn his soft, water-loving polymer in a pliable lens. In his kitchen he built a mold the shape of a lens, mounted it on a child's erector set, then used the motor of an old phonograph to spin the mold. Witcherle sprayed polymer on the spinning shape, and created a thin, perfectly formed contact lens.

By early 1962 he and his wife, a doctor, had produced five thousand of them. Wichterle traveled around the world distributing handfuls of the lenses to interested ophthalmologists and optometrists. "The reaction was unanimous," he later said, "They were a joke, an interesting subject, but without any wider application." Nothing much happened until he demonstrated his contact lens for some U.S. patent lawyers touring Prague. He recalled: "I took a lens out of my eye, threw it on the floor, stepped on it, then washed it with my mouth and put it back in my eye." The lawyers were impressed enough to buy the patent rights.

The Czech government received the lion's share of the money; Wichterle got almost none. He took the loss philosophically, saying later: "I would have had problems with what to do with such an amount of money." Instead he just kept inventing - earning hundreds of patents on biomedical plastics and synthetic fibers. And also Wichterle earned huge academic distinction and international recognition for inventing the contact lens - although he always wore glasses.

Matches

HERBERT SPENCER, a great philosopher in the 19TH century, once pontificated on matches. He said that "the [friction match was the] greatest boon and blessing that had come to mankind in the nineteenth century." Indeed until Spencer's time keeping a flame alive was a central problem of life.

If it went out the best way to make a fire was the tinder box. That is a metal box containing a piece of flint, a block of metal, and the tinder - a dry flammable material like charred linen, dried fungi, or feathers. The hopeful fire builder struck the flint on the metal to pare off a tiny flaming fragment, which fell upon the tinder, igniting it. Charles Dickens claimed that by this method on a damp day one might get a light in half an hour "with luck."

For most people this tinder box was the only way to make a flame until the early 1800s when John Walker, an English pharmacist, invented the friction match. He covered a sliver of wood with some chemicals so that if the head were nipped between folds of fine sandpaper it ignited - more often than not, as long as the head didn't fall off. But the matches were expensive, unreliable and difficult to strike. And they made a noxious puff of smoke that smelled like rotten eggs. One brand called Lucifers—from the Latin for "bringer of light"—came with a warning: "If possible, avoid inhaling gas ... Persons whose lungs are delicate should by no means use the Lucifers."

The next big step for matches and humankind was a simple chemical: Phosphorus. It had been discovered centuries before matches; its highly ignitable, you can light it just by striking it on something. But it was expensive because it was made from animal urine, it took barrels just to make a single ounce of ignitable phosphorus. So, for decades the rich would carry a small box of phosphorus and amuse their friends with its ignition. But then phosphorus was discovered in the bones of animals—and so could be made cheaply.

Yet it still presented problems: Used on matches it was too easy to ignite—it exploded violently. A Swede solved this explosion problem by inventing safety matches. He put the explosive phosphorus in the sandpaper outside of the box. When dragged along the box the chemicals on the match's head reacted with the phosphorus in the sandpaper to make a flame.

A flame that was considered miraculous at the time, and is now mundane. We can even mark the year when striking a match became commonplace and was no longer an awe inspiring event: In 1889 the first matchbooks appeared with advertisements on them.

Demolition

I T DOESN'T SURPRISE ME that we live in a disposable society, its the scale that startles me.

I'm thinking of the recent demolition of two sports stadiums: Both Three Rivers Stadium in Pittsburgh and the Kingdome in Seattle have disappeared. The Kingdome seemed young as structures go: It was only twenty four years old. And successful too: It had seen seventy three million visitors who watched the Mariners, Monster Truck shows and even religious rallies.

Some days I think about how we dispose of huge buildings and wonder whether the Romans would have got rid of their Colosseum. Perhaps they would have. I don't think human nature has really changed much over the centuries—most likely the Romans just didn't have the technology to demolish the Colosseum. To get rid of a superstructure like Seattle's Kingdome takes an incredible amount of knowledge about structures, mathematics and explosives.

The Kingdome, for example, presented a special problem to the demolition team. The dome is based on the Roman arch—in a sense the dome is one large solid arch—which is a very, very strong thing. Its held up only by supports around it edges. Now it would seem that all you have to do is knock out these supports and let the dome fall. But if you did this the Kingdome would shake the ground and destroy nearby buildings. Mark Loizeaux, head of the demolition team, said that the trick is to let it fall gently.

To do this Loizeaux first had teams remove the Kingdome's seats and bleachers, then fill it with twelve thousand cubic yards of concrete rubble to cushion the fall of the dome. Next his explosive experts drilled 5,500 holes throughout the dome, filling them with over four thousand pounds of explosives. Loizeaux's goal was to time the explosions to cut the dome into six pie-shaped pieces—each falling before the other, thus avoiding the huge shook of the entire dome crashing to the earth. His goal was to leave unharmed buildings as close as one hundred feet.

Just before the demolition of Seattle's Kingdome, on a Sunday morning in March, Mark Loizeaux and his team, as is their custom, took a moment to pray. Next they started a ten second countdown, followed by a wait of five seconds to let crews shout a final warning. Then they pressed the button. The explosions took out the dome in six pie-shaped slices in sequence, pulling the structure inward—creating an explosion equal to an earthquake of two point three on the Richter scale. In total it took 16.8 seconds to fall—they'd guessed it would take 17.8, this is very precise work.

The Kingdome, home to a billion happy memories, was now a pile of rubble only thirty-two feet tall. Surely the ultimate monument to our disposable culture.

Bose Wave Radio

M Y WIFE RECENTLY BOUGHT a Bose Wave radio. For such a tiny device—its only the size of a clock radio—it makes a huge sound. Unlike most brand names there truly is a person behind this one; in fact Amar Bose still heads the Bose Corporation.

The wave radio began in the 1960s when Bose went shopping for speakers. Being very attuned to sound, he'd studied the violin for years, the systems of the time didn't live up to his expectations. His frustration in finding speakers propelled him into a career in psychoacoustics: That is, correlating how sound is produced electronically with the way people perceive it.

Having just finished his doctorate in electrical engineering, he was sure he could build better speakers. So he started the Corporation, based on a motto of "better sound through research." The first speakers he developed were technically excellent, but they failed: They were too large and complex for most people. Bose said this failure told him "people want something that gives you the full benefit of a hi-fi but [is] as simple as a refrigerator."

Meeting this goal took him and his co-workers fifteen years and fourteen million dollars. They wanted to make a compact stereo system that produced all frequencies of sound. Normal speakers can produce only a narrow swatch of sound in the middle of the range where the human voice falls. Usually voices sound nasal, or tinny, because the bass frequencies are missing. Bose wanted all of these frequencies—high and low—in his tiny speakers so he could create a full, rich sound. He knew that a pipe organ makes incredibly rich sounds—and that's essentially what he put into his Wave Radios.

How, though, to fit an organ pipe into a tiny box. Bose solved this problem by designing a complex, twisting sound tube that winds around for 34 inches back and forth in the body of the radio—although the device is not much bigger than a clock radio. If you peer inside it looks a bit like an inner ear. This isn't as simple as it sounds —they key is how to fold the tube to let it make all sorts of sound. According to Bose "the math alone could fill a wall." And this tube must be perfectly sealed and its dimensions must be extremely accurate.

Amar Bose attributes his success to owning his own company: He controls two-thirds, the rest held by private investors. It isn't public, you couldn't buy a share if you wanted to. "Public companies," he says, "are by nature short-term oriented. They have to make their financial statements look good every 90 days." And that, he adds, is "a formula for eventual loss." He might well know what he's talking about: the name Bose is now the world's largest audio brand.

Stay-on Tab

I JUST INSTALLED A pop can crusher in my home and it saddens me every time I use it.

To me a pop can is a thing of incredible beauty and elegance, even the tab used to open the can is an engineering masterpiece. Recently I opened thirty cans of cheap root beer just to see how this tab worked.

As I studied the tabs, in a room reeking of root beer, I got more and more interested. So I called its inventor. Dan Cudzik begin by telling me it took him five years to invent the stay-on-tab. I said to him "Wasn't that a lot of work for a piece of metal?" Dan said: "It wasn't a piece of metal, it was a whole concept." Dan explained that he was racing every can company in America to invent the stay on tab.

The removable tabs of the sixties where an environmental hazard: They littered beaches and also children would sometimes eat them. This drove Dan Cudzik to find a way to keep the tab on the can. He faced this problem: The can has a perforated disk of metal which must be snapped out of the opening.

When Dan tried to use the tab as a lever to do this it either snapped off, or required him to build a huge tab—one too large to be economically manufactured. Dan became so obsessed with this problem that he would wake in the middle of the night, his mind filled with ideas.

To help him think he would climb into his car and drive 110 miles to the mountains nearby, then drive straight home arriving in the morning as his family was rising. One evening, after nearly five years of thinking, Dan was watching TV with his kids and his wife. And the whole concept hit him "like a ton of bricks." He rushed up to his kitchen and began to sketch.

The key, Dan realized, was to use the lever—the tab we pull—to widen the opening in the top of the can and let the perforated disk fall into the can, instead of pushing it through the opening. The next day Dan built a working model from cardboard, aluminum foil, and tape. It was ten times the size of a normal can, and with its swinging inner lid, it looked a lot like a toilet seat. Dan tested it again and again—and every time it worked fine.

Today Dan Cudzik's stay-on-tab is used on every one of the 100 billion pop cans that are made a year. And now you can know see why I'm saddened when I take one of Dan's tiny engineering masterpieces and destroy it in my can crusher.

Cornstarch Packing Peanuts

M Y WIFE AND I GOT A package in the mail that fascinated me. I don't even recall its contents, because I was taken with the the green packing peanuts used to protect whatever it was from damage. As I scooped up the pellets to toss them in the trash, my wife said, with a very knowing voice, "Just toss them on the compost pile." What! Plastic in the compost? No. She showed me a slip of paper that explained: There was no "plastic or polluting gases" used to make these peanuts; they were made of cornstarch.

Toss them on your compost pile or spread them on your lawn and with a bit of water they'll dissolve in minutes. These cornstarch packing peanuts are part of a movement called "green engineering."

It's a design philosophy where the environment is explicitly considered from the beginning: A goal is to find processes and products which are feasible and economical while minimizing pollution at the very beginning. These cornstarch packing peanuts are the work of food engineer Bill Stoll.

He grew up in a small Iowa farming town and attributes his creativity to this upbringing. "Farmers," he recalls, "could fix anything." Watching them do this gave Stoll a love for entrepreneurial ventures. He claims it's "The most exciting thing a person could do -- and the scariest. You challenge every creative bone in your body -- like jumping off a ledge with a bungee cord."

So he devoted his career to consulting with entrepreneurs in the food industry. In 1992, near the end of his career, Bill Stoll had lunch in a St. Paul restaurant with a client whose company used popcorn for packaging. The owner was looking for a way to pop bigger batches and he wanted to pick Stoll's mind. This question made Stoll recall sounds he'd heard as a kid.

In a factory he'd seen cereal being prepared in huge pressure cookers. When the clamp holding the top was knocked away by a sledge hammer, the lid flew open and the grain exploded as if from a cannon. Stoll knew that a similar method is used to make puffed snack food. So he came up with the idea of making something like corn curls for packaging. The result: biodegradable packing peanuts.

Now that I've seen these I see cornstarch everywhere. I've heard of biodegradable cornstarch cutlery, it dissolves in a day. Not a bad deal considering a plastic fork is typically used for three minutes, and then sits at least a hundred years in a landfill. And I've heard rumors that a company is developing a way to use cornstarch to make upholstery. Who knows where this green engineering revolution will end.

Claude Shannon

IN THE 1960S A POPULAR PHRASE was Marshall McLuhan's "the medium is the message." Yet at least twenty years before McLuhan's aphorism an electrical engineer had disproved it.

The engineer, Claude Shannon, who died recently, used mathematics to separate the medium from the message. Today we hear talk of bandwidth, and digital phones, and fast lines to connect to the internet.

It all began in 1948 when Shannon showed that all messages could be represented by just two numbers—zero and one—the binary form of data which is so popular now that we all have computers. He was even the first to use the word "bit" in print. It's a contraction of the first two letters of "binary" and the last letter of "digit."

He used very eloquent and even beautiful mathematics to show how many of these bits were needed to send information through a channel.

By channel he meant anything: a copper wire, smoke signals, a fiber optic cable, and even the electromagnetic waves that travel through the air to your television or radio. By breaking a message into these bits he could distinguish the message from the medium and allow other engineers to focus on the message itself, without being concerned if it were, say, voice, data or telegraph. Using his theories they could design a channel of just the right size to transmit a message without errors.

For a man who made it possible for information to flow around the world Claude Shannon was surprisingly reclusive. He worked alone at Bell Labs in New Jersey, usually keeping his door shut. Emerging only at night to ride his unicycle down the halls, occasionally even juggling while on the bike. At times he'd appear with some bizarre invention: a two-seated unicycle, juggling robots, a motorized pogo-stick, chess-playing machines, or a rocket-powered Frisbee. He devoted much of his later years to developing a mechanical mouse. Fitted with copper whiskers, a magnet and wheels. It could find its way through a maze. Most intriguing, though, to his colleagues was a mathematical system he designed to analyze the stock market. He made a handsome profit, but the details of his methods remain unpublished.

In a sad irony Claude Shannon, who brought information to the world, was unable to communicate toward the end of his life. He suffered from Alzheimer's disease. But his theories live on. They've allowed engineers to bring us digital phones, high-speed internet lines, and the flawless audio and video we expect of compact discs and DVDs.

Although Claude Shannon helped connect the globe he still didn't share everything with us. He took to his grave his design for a rocket-powered Frisbee.

Ping Putter

SPRING IS HERE AND IF YOU listen carefully you'll hear the sound "ping", followed perhaps by swearing. It's the sound of golfers enjoying their game. That "ping" is the sound of one of the most revolutionary putters of all time called, of course, the Ping Putter. It's the invention of engineer Karsten Solheim.

He worked for years designing jet fighters and missile guidance systems at General Electric, before taking up golf at age 42. His first time, he recalls, "I hit the ball ten times, and then walked away ...." He didn't even make it to the first hole. Solheim had became frustrated while practicing putting. Instead of quitting he just designed his own club. Solheim noticed that if a player didn't strike the ball perfectly with a normal putter, the head would twist destroying the shot. So he moved the weight of the club head to its outer edge making it more stable, and letting the player putt straighter. This is called "perimeter weighting" and has been ranked as one of the top five breakthroughs in golf equipment.

At night, after work, Solheim kept busy in his garage refining his revolutionary design. He first worked with two Popsicle sticks glued to two sugar cubes attached to a shaft to find the correct weighing. By 1959 he founded the Karsten Manufacturing Corporation to produce his putters. With a one thousand one hundred dollar bank loan -- his only debt ever -- he went to work in his garage in 1961.

Six years later he quit GE to make clubs full time. He called his new club the "Ping Putter" because of the sound it made when striking the ball. He spared no expense or effort in making nearly perfect clubs; for example, he pioneered a special type of metal casting to give the iron in his putters great consistency. Yet, Solheim had great difficulty persuading golfers to use a putter which looked so unfamiliar.

Promoting his designs at professional tournaments, he was treated as an eccentric to be gently humored. But before long his putters could be seen in more bags than not. The turning point came in 1967, when Julius Boros, a winner of three majors, won the Phoenix Open using one of Solheim's putters. "The putter," said Boros, "looks like a bunch of nuts and bolts welded together, but the ball goes in the hole." Solheim's relentless pursuit of perfection revolutionized club making.

He built clubs of such quality that he forced the rest of the industry to upgrade its standards. So far ahead of the curve was Karsten Solheim that at times his innovative clubs were banned from official play, until golfers demanded they be allowed. It's probably no coincidence that the boom in golf equipment started about a decade after Solheim founded Ping.

Cell Phones

I'M FASCINATED BY EXPERIENCES made possible by technology —things never felt or seen before the 20TH century. I like feeling the incredible acceleration of a jet. I like seeing rows of mass produced objects—miles of things identical in shape, size, and color. It doesn't matter to me if its a warehouse of bathtubs or ball bearings, I get a thrill out of seeing them repeat. Not too long ago technology gave me a new experience, a humbling one, which never could have happened even twenty years ago.

It happened before on-line bookstores were popular. Back in those days you had the bookstore order a book and then you stopped by and picked it up. It's something I did so often I couldn't even keep track of all my orders.

One day, in this ancient time, I went to the bookstore to browse, first stopping at the coffee shop to get a bagel.

Now I'm standing in line, waiting to order and my blood sugar is extraordinary low. And I'm extraordinarily grumpy because the guy behind the corner is talking on the telephone. Now I believe if you're behind a counter you should provide service especially to me when my blood sugar is low. I was about to find his manager when I noticed he had a book in front of him. He opened the book and said "The book you ordered is in, please pick it up at your convenience."

I realize now this isn't his fault, it's the managements: They shouldn't be having the coffee guy call people about books especially when I need a bagel.

Finally he does turn to me and we do this "Green Eggs and Ham" thing. I say I want a bagel. He says would you like cream cheese? No, I would not like cream cheese. Would you like it cut? No, I would not like it cut. Would you like it on a plate? No, I would not like it on a plate I would like you to hand it directly to me so I can pound it on the escalator before I faint from low blood sugar. Then I got in line to buy my books and here is where technology enters. This story would not be possible without it.

I pulled out my cell phone and called my answering machine. I'm always calling my machine because my friends can never get their lives in order; so I just say "Look, call my machine, I'll check it all the time." And what do you think I hear? I hear the bagel man say: "Mr. Hammack, the book your ordered is in. Please pick it up at your convenience."

I walked back upstairs and quietly got my book.

Moen Faucets

M Y WIFE AND I HAD A new faucet installed, the kind where the water flow is controlled by a single handle. This single-handled faucet has been listed as one of the top 100 mass-produced objects of the century—right along side Henry Ford's Model-T. The impetus to invent this faucet came from an accident.

In 1937 Alfred Moen worked part time in an auto garage to earn money for engineering school. Late one night, tired and ready to go home, Moen stopped to wash his hands in a sink with the usual two handled faucet. Moen turned the wrong handle, creating a burst of hot water, and burned his hands. This gave the budding mechanical engineer his lifetime passion: the perfect faucet.

What Moen confronted was a common problem of any human/machine interface: How to communicate, instantly, to the user the way to control the device. Usually a faucet designer takes the easy way by giving us clear pointers to the hot and cold water—the two knobs of a typical faucet—but that isn't what we want to control: As Moen realized, we want to control the volumetric flow and the temperature. And that's the problem Al Moen tackled.

"The more I thought about it," Moen said, "the more I was convinced that a single-handle mixing faucet was the answer, so I began to make some drawings." He first designed a complicated faucet with a cam to control the flow of hot and cold water, but it was too complex to be manufactured cheaply. As Moen refined his faucet World War II broke out, making metal scarce, so he went to work as a tool designer in Seattle, then served in the Navy.

But he never abandoned his vision for a perfect faucet, and by 1947, he'd reinvented it and interested a manufacturer. In that year they sold their first 250 faucets to a San Francisco plumbing supplier for twelve dollars a piece. Moen's invention transformed the industry to the point where today more than 70% of kitchen faucets sold in the United States are now Moen's one-handled variety.

Oddly, he never owned the company that still bears his name, preferring instead a behind-the-scenes role. He spent his life in the research and development section compulsively refining his faucet, and even inventing a special valve to prevent shower shock; the rude surge of hot water caused when someone flushes a toilet while another person is in the shower. So, appropriately, Alfred Moen listed only one occupation on his business card: "Inventor." When his son asked him, near the end of his life, if he were disappointed at not having been elected to the National Inventors Hall of Fame, he replied: "No, I didn't invent anything great. I didn't invent penicillin"—just a faucet with a single handle.

Margaret Knight & Paper bags

THE NEXT TIME YOU'RE AT THE grocery store and they say "paper or plastic?" I want you to choose paper, and when you get home examine the bottom of the bag. That square, flat bottom—called a satchel bottom—is the work of Margaret Knight.

Over a hundred years ago she invented a machine to make these flat bottomed bags, and it revolutionized retail sales in American. The large New York department stores, Macy's and Lord & Taylor's, realized they could use Knight's cheap, but sturdy, flat-bottomed bags to allow a clerk to quickly package a purchase and move on to the next customer, instead of taking the time to wrap a parcel with paper and twine.

Knight's first invention was at age twelve. She visited a cotton mill, to see her brothers who worked there, and saw an accident: A spindle flew loose and wounded a girl. This spurred Knight to figure out a way to make the mills automatically stop, thus preventing further accidents. After this invention Margaret Knight doesn't appear in any public record until her first patents twenty years later, when she was thirty-two.

Two years after the Civil War she went to work for a paper bag manufacturer. While there she invented an ingenious device for taking a roll of paper and converting it into flat-bottom bags.

The machine is an incredibly complex and clever thing: It cuts a strip of paper from the roll, glues it into a tube, cuts the ends and folds them neatly and quickly into a flat bottom. Margaret Knight built a wooden prototype of her machine and made thousands of trial bags. As her prototype churned away, a man visited the factory, studied the machine, and before she could patent it, stole the idea.

Knight sued and then spent the incredible sum, at the time, of $100 a day plus expenses for sixteen days of deposition of herself and key witnesses. Due to her careful notes, diary entries, samples and expertise the court ruled in her favor.

After the suit Margaret Knight continued inventing—working on machines for making shoes, and rotary engines. By the end of her life she'd patented over 22 inventions, and even got a decoration from Queen Victoria in 1871 for her paper bag machine.

Still she felt she could have done more. A year or two before she died, she told an interviewer: "I'm not surprised by what I've done. I'm only sorry I couldn't have had as good a chance as a boy, and have been put to my trade regularly."

Digital data

A FRIEND OF MINE RECENTLY bought a digital camera to record the life of his newly born daughter. His goal is to create a permanent record of her life, something she can enjoy years from now. But how permanent will this be? And how truthful a record will it be for historians of the future?

The answer is that digital recordings will likely be neither permanent nor trustworthy. To see why note how you view digital photos or listen to music. To see or hear them you need a special machine—a CD player or a personal computer. I know how common those machines are today, but the key word is today.

Visit the National Archives in Washington, D.C. and you'll see the problem. They have a Department of Special Media Preservation, nearly a museum of obsolete technology. Here engineers try to listen to early records made of glass, or audio tape made of thin wire. Even worse, they are faced with millions of documents created and stored on now obsolete computer systems. The documents are unreadable unless they rebuild these ancient computers. Will your computer, CD player or digital camera still work years from now? Probably not.

Think of data created only fifteen years ago; it was recorded on five and a half inch floppies, which most computers can no longer even read now. And what about the data itself: Won't a CD or a digital tape always be around? The answer seems to be no. Some manufacturers say CDs will last ten to fifteen years, some experts think it will be about forty years, but this is for the CDs you buy at the music store. The CDs you might burn at home are made differently and their lifetimes are even shorter.

And what of digital video tapes? Likely even worse: They must be stored correctly to survive heat and humidity. Even if all of the tapes and disks lasted and we had computers to read them, how accurate a record would they be?

Today with a home computer you can easily change photos—moving bits and pieces here and there to create a fake photo no one can detect. Digital makes things easy to tamper with, unlike traditional photographs, which took great skill to be faked. So as a society we may either leave no track record—all our history tied up in obsolete and decaying technology, and what is left may not be trusted.

What is the solution? Well, I visited our university archivist and asked him. His answer: Etch it in steel or stone, bake letters into clay, or even use paper. If you want to be remembered write a book and have it published. Thousands of copies of it will exist in libraries around the world; and it will be accessible to anyone without equipment.

Bolts

RECENTLY I WATCHED A man trust his life to the threads of a bolt. His name is Noah Bigwood, he's a world class climber, who can scamper up a rock face that looks like a shear vertical slab. I met him while he was guiding my wife and some of our friends on rock climbs in Moab, Utah, although he didn't guide me because dangling from a bolt on a cliff isn't my idea of fun.

To guide climbers, Noah first works his way to the top of the slab where there is a bolt with a metal loop attached. He then slings a rope through this, descends the rock, and uses this rope to belay the other climbers, like my wife. The safety of this whole enterprise depends on the threads of a three-eighths inch diameter, four inch long stainless steel bolt. More specifically it depends on being able to buy a mass produced, high quality bolt.

This fact makes recreation climbing, then, a hobby made possible by the Industrial Revolution of the 19TH century. We can still see the fruits of 19TH century thread design in today's bolts. Go to a hardware store and pick out a large bolt and look at the bottom of its thread. You'll notice that it is flat and not pointed like a Vee. It is this flatness at the bottom that is a key to mass producing a bolt.

Mass production was partly driven by the needs of railroads. As they spread westward locomotives were often repaired hundreds of miles from the central shop, thus the bolts used for repair had to be uniform for interchangeability. To be uniform required a standard system. At the start of the Industrial Revolution the English system prevailed. These bolts featured threads that came to a sharp point at the bottom. These bolts worked just fine, yet they could not meet America's need to fill the world with mass produced objects. This simple English screw was too complicated.

To make the very sharp vee-notch of the English screw took an expert machinist using five different machines. In Britain this worked well because there was a surplus of labor; but in America, which was expanding in leaps and bounds, there wasn't enough highly skilled labor available. In response to this William Sellers, America's pre-eminent tool-maker, designed a bolt thread able to be made not by experts, but by "all good practical mechanics."

Sellers' improvement was extremely simple: He flattened the bottom of the Vee thread. This simple redesign allowed the bolt to be made easily by any component machinist.

That flat cut at the bottom is engineering at its best: It isn't some fundamental scientific principle, it isn't even artistic. Its just good, practical engineering that makes possible just about every mass produced object—including the bolts used for climbing mountains.

Composting Toilets

IN VANCOUVER, BRITISH COLUMBIA there is a building that people come from around the world to see. For some the trip is nearly like visiting a shrine, others are driven by mere curiosity. The building is the C.K. Choi Institute for Asian Research at the University of British Columbia. It is a striking building of purplish brick, hidden behind rows of young ginko trees. And its roof is made of five magnificent silver arches that look like the billowing sails of a ship. Yet people visit just to see its toilets.

It features flushless, waterless composting toilets. They are completely disconnected from the sewer system, instead the run off is used as fertilizer for the surrounding plants. This might sound like the pit toilet from your camping days, but it's actually a very high-tech cousin featuring rotating tines, temperature and moisture probes and electronic control systems.

What goes into the toilets passes through a system of five trays at the bottom of the toilets's 14-inch stainless steel chutes. Maintenance staff toss in wood chips to aid composting, and red wiggler worms burrow through the muck to turn human digestive by-produce into topsoil.

Any leftover liquids, called composting tea, join the building's used water in a trench along the side of the building, where it's cleaned by micro-organisms on plant roots before being stored for summer irrigation. The system saves more than five hundred gallons of water a day, and eliminates a huge load on the sewage system.

These toilets in the C.K. Choi Institute represent a movement toward a sustainable building. Although the toilets are the most interesting aspect, this idea of sustainability was used throughout the building. For example, construction workers made the structure's heavy timber frame with lumber recovered from previously demolished campus buildings. And they reused red brick cladding from Vancouver's streets. The building is very narrow so that each room gets the maximal amount of sunlight. Sensors measure the amount of sunlight in a room, and if it's adequate the electrical lights are automatically dimmed. This makes the building use less than half the amount of lighting of a conventional building.

The five arches of the peaked roof aid the natural ventilation: Air from vents in the lower floors rises to the top of the giant peaks continually flushing fresh air though out the building, and reducing the air conditioning load.

This movement toward constructing sustainable buildings is spreading to homes. People are even retrofitting existing homes to be more green.

You could, for example, put a composting toilet in your home. There are many books and web sites with workshop plans showing all the how and why's of these toilets. You might first, though, wish to make a pilgrimage to the toilets in Vancouver's C.K. Choi Building, because, as one home user warns, "living with a composting toilet is a major [life] decision."

Science & the Declaration

EVERY FOURTH OF JULY I read the *Declaration of Independence*. I do this partly because I enjoy its eloquent phrases, partly because its lofty sentiments fill me with historical pride, but mostly I reread it to be a better citizen. It is, after all, the founding document.

Over the years I've detected, with my engineers eye, an unmistakable trace of science and math in the Declaration. It reads like a geometric proof with its "laws of nature" and its truths held to be "self-evident" like axioms. It first lays down axioms like "All men are created equal" and then derives, if you will, an indictment against King George III.

I've learned that there is more than an echo of scientific reasoning in the Declaration, phrases like "laws of nature" had deep meaning for the Founders. Thomas Jefferson, John Adams and Benjamin Franklin, all members of the committee that wrote the Declaration, used science as a source for metaphors. They believed it to be the supreme expression of human reason. For no Founder was science more important than Jefferson, the Declaration's main author.

In a letter he revealed, "Nature intended me for the tranquil pursuits of science, by rendering them my supreme delight." He filled his writings with discussions of plows, air pumps, compasses, canal locks, balloons and steam power. He stocked his library at Monticello with books on every aspect of science and technology. None more important to him than Euclid's geometry and Isaac Newton's great works on physics.

At Monticello Jefferson had a picture gallery of intellectual giants. He assigned a high place to three portraits, one of which was Isaac Newton. He esteemed Newton as one of the greatest minds the world had produced. Jefferson had both the Latin original and English translations of Newton's *Principia* in his library; he is surely our only president who's actually read it. So, when Jefferson opened the *Declaration of Independence* by asking the American people to assume the powers of the Earth "to which the Laws of Nature .... entitle[d] them." he meant more than natural law—the supreme moral law known to humankind through reason.

To Jefferson and the other Founders, the words "Laws of Nature" had a deeper resonance: They evoked a picture of Newton's laws of motion, of the universe as one great harmonious order obeying mathematical laws. A world in which it was natural that humans have inalienable rights.

How does all this help me today to be a better citizen? Well, in hindsight, the Founders too glibly made a leap from the laws of gravity to the laws of human interaction. In reading the *Declaration* one can easily forget that the dignity and "the rights of man" are neither a self-evident axiom, nor an inalienable right, but instead a hard-earned acquisition that we must continually work to keep.

Roller Coasters

I HAVE GREAT FAITH IN MY fellow engineers, but there is one piece of technology used by three hundred million people, which I am never going to try. It's a roller coaster.

They terrify me, although engineers design them to be so safe that insurance companies worry more about sprained ankles from Merry-go-rounds. The engineer's goal is to use gravity and acceleration to confuse, and apparently delight, a rider.

The main thrill is the *g*-forces—the changes in weight from acceleration. When the cars take off a rider feels about two and a half gees, nearly what the astronauts feel when the space shuttle launches. By understanding these *g*-forces, engineers have been able to enhance riders' thrills, and prevent their deaths. An early coaster called the "Flip-Flap" had the first loop-the-loop, which is a complete circle that turns riders upside down. Its designers understood *g*-forces so poorly that as the cars went around the loop they exerted twelve gees on the riders. Fighter jet pilots black out usually at ten gees. This coaster occasionally snapped riders' necks. Today's engineers use computers to calculate the forces on every section of the track.

It isn't, though, only technical skill that makes a good roller coaster. Knowing some psychology also helps. They've learned over the years that it's best to have a slow ascent on the first hill, then dangle people at the top before the great plunge. On that plunge, engineers try to make cars go as straight down as possible and appear to curl underneath, giving riders the impression that they're about to jump off the track. A good roller coaster engineer also makes use of the supporting structure. Sending a car close to a column gives the impression of speed, and a classic coaster trick is the *fine del cap*— Latin for *end of the head*. By this they mean shooting the cars toward a horizontal beam, then ducking at just the last moment.

With these principles in mind, here are tips from me, a non-rider, but a knowledgeable engineer on how to make your ride more scary, and I gather, more fun. The front car feels fastest. Close your eyes briefly during the ride it will help confuse you. Also look backward to get dizzy. Lift you feet as the car crests a hill—you'll feel weightless. And ride after a rain because the tracks are slippery and the coaster is faster.

And another tip is to keep coming back, because they're always improving the rides. For example, some new coasters no longer start on a hill, instead they are driven by motors used to power rockets. With these motors, roller coasters blast off with accelerations of 4.5 gs. Myself, I'll be on the ground, holding cotton candy, and feeling only one *g*.

IETF & Domain Names

YESTERDAY, WHILE WEB SURFING, I got to wondering: Who controls the Internet? I'd always had this image of a giant, unorganized free-for all. Yet, it turns out the Internet is a highly organized and regulated affair. Three non-profit organizations control it.

One of these is the IETF, the Internet Engineering Task Force, that sets the technical standards. And who exactly is the IETF? Well, it could be you or me. As their web page says "no cards, no dues, no secret handshakes." Just sign up, attend a meeting and propose something. They'll debate your suggestion and then see if there is a "rough consensus" to adopt it. By that they mean something beyond a simple majority, but short of unanimity. When the committee votes it's often done by humming. That way no one can tell who is for or against a proposal, but that rough consensus can be determined. Somehow I find it a bit disconcerting to know the information superhighway is being built by humming.

If you want real power, though, you'll need to join a committee where they don't hum. Namely, ICANN, the Internet Corporation for Assigned Names and Numbers. Like the IETF this is one of the big three that run the Internet. They control who gets what web site name, called a domain name. When you purchase the name, ICANN registers it on one of their thirteen computers, called root servers. Then when you type in, for example, *www.ebay.com*, one of their servers tells your computer where to find the web site.

This gives the ICANN committee immense power because having the proper name can be life or death for an on-line business. Yet it's run by only seven full-time staff, directed by a nineteen member board of directors. Just like all the non-profits that run the Internet you could be a member of this board. They're selected by direct election and anyone who is over sixteen years old and who has an e-mail and physical address can be nominated.

If you serve on this committee be prepared for controversy. As the Internet expands exponentially, ICANN is running out of domain names. Currently they only allow ones that end in .com, .org, or .net, and for these names the most recognizable English words have already been claimed. After much debate, and critic's cries that ICANN is moving at less than Internet speed, they've allowed seven new endings. Among them .biz for business.

This .biz ending has started another gold rush for domain names. In the past if you owned a dot com name desired by a large company or a Hollywood star you could sell it for lots of cash. This alarmed me, so I logged on recently and was relieved to get the domain name of my choice. Feel free to visit *www.billlhammack.com*

For curious listeners the third non-profit running the internet is the w3c—the World Wide Web Consortium. They set the standards and protocols for the web.

Ice Cream

RECENTLY I TOSSED OUT MY ice cream maker. I've tried for years to make great ice cream; but it was never as good as what I could buy. So, I took a look at how ice cream is manufactured. I learned it's a very tricky business.

To the food engineer it's just air bubbles, oil globs and ice crystals suspended in water, but the key to engineering ice cream is getting all these bubbles, globs and crystals to be the right size. The oil comes from milk fat, which, of course, doesn't mix readily with the water. So the first step is a device that shatters the fat globules so they'll mix with the water, otherwise the cream will float to the top. Next, air is pumped in.

Ice cream can be up to fifty percent by weight air, although by law a half gallon must weight at least two pounds, two ounces. Adding in the right amount of air can make the difference between mediocre and excellent ice cream.

Too much air insulates the ice cream, making it melt slowly in our mouths and ruining its flavor. As it melts the chemicals containing the flavor are actually boiled on our tongues. The tropical orchid vanilla, for example, contains dozens of flavors. Each has a different boiling point that makes the vanilla play out in a certain time sequence on our taste buds. The more air then, the slower the melting and the less rich the flavor.

So, when you're in the supermarket checking out ice cream what should you look for? Find the highest density ice cream; that is, the heaviest half gallon. Most ice creams weigh close to the legal limit of two pounds, but premium brands can be nearly twice as heavy.

After you buy it I suggest you rush home. It is imperative to keep the ice cream at a constant temperature, otherwise it loses texture. If the ice crystals are bigger than one-thousandth of an inch, the touch receptors in the roofs of our mouths detect them. So, when you get that ice cream home, toss it in the freezer, and keep it closed. Opening the door makes the freezer heat up only slightly, but enough to melt the ice crystals. Then when the door is closed the water freezes, but this time freezing into larger crystals that make the ice cream crunchy.

One more ice cream tip for you: If you want excellent ice cream, go to New Zealand. They rank number one, per capita, in ice cream consumption. Each person eats about seven gallons a year. Their snow-fed rivers, clean air, sunshine and year-round grazing on rolling pastures lets their cows make milk that produces some of the best ice cream you'll ever taste.

Superglue

A T A PICNIC YESTERDAY SOME friends and I got to talking about what makes my tiny cell phone possible. One person suggested the ceramic used to make the transmitter, another spoke of the miracle of the microchip, but curmudgeonly I contented it was glue.

Glue is such a humble thing, yet think of our world without it. No grocery bags, envelopes, books, magazines, paper cups or cardboard boxes. And that's only the low tech end. It also keeps together our microwaves, refrigerators, cars and jets.

Glue also gives our era its sense of style. We associate sleekness with technological advancement, something made possible by glue. Examine any 19TH century object and you'll find all sorts of fasteners: bolts, nuts, rivets, pins, staples, nails, screws, stitches, straps, and even bent flaps of tin. Pick up a cell phone and you'll find only a smooth, sleek surface; the pieces of the phone are held together by a type of superglue.

Two Kodak researchers discovered the glue by accident in the 1940s. As part of the war effort they searched for clear plastic substances for gun sights. They mixed up a batch of what we now call superglue, tossed it between two expensive prisms and measured how much light passed through it. Of course, they found their prisms were stuck together and ruined. They uncovered, though, a key property of superglue, one that makes it so useful today: Superglue works best when applied as a thin layer. This makes superglue unique. For all other glues, a thicker application holds better, but superglue works best with only a tiny amount.

This property makes things like my cell phone look sleek and modern, no globs sticking out. Not only does this thin layer give our material world its look, it also allows superglue to help us understand our past.

It's an essential tool used by paleontologists to preserve fossils. Most fossils burst apart as soon as they're moved from the dig site. In the past, paleontologists covered fossils with Elmer's glue or shellac, but these glues don't penetrate. They reinforced the surface of the fossil, but left the inside weak.

As one paleontologist says, "It's like adding strength by painting a barn." Today they use superglue, which quickly penetrates tiny crevices, then hardens to strengthen the fossil.

What's next for superglue? Us. The FDA recently approved replacing traditional surgical sutures with a special type of superglue.

Color Film & Determinism

I THINK THE MOST COMMON IMAGE of technology is a great whirlwind of innovation that sweeps through our lives, creating blessings and havoc. This view is easy to hold in this day and age when there is a headline every day with some new breakthrough, but it is only half true, and because of this, dangerous.

Its error lies in treating a technological breakthrough or object as culturally neutral. In reality the technical aspects cannot be separated from their social context. The values and world views, the intelligence and stupidity, the biases and vested interests of those who design a technology are embedded in the technology itself. You can see this every time you go to a movie.

On screen the colors appear to be a representation of reality, but are not at all. They were optimized to best reproduce caucasian skin tones. This came about because it's impossible to faithfully reproduce colors on film and to do it cheaply. Instead, engineers developed ways to simulate a full color picture using three colors—blue, red and yellow. But these three colors cannot create a full spectrum of pure color.

For example, if you use them to create white, it is tinged with magenta. You can restore full whiteness by fiddling with the mix of the three primary colors, but only at the cost of distorting all other tones.

To find the best color balance, the engineers developing film stocks used a series of prints of a young woman. They varied the color balance in the prints from too red or yellow to too blue, and from too green to too pink. They submitted these to judges, who chose the best reproduction. They rejected the print that matched the real colors in the woman's face, and choose as best, the prints where she was quite pale. Thus they optimized film for very light Caucasian skin over all other skin tones.

This isn't to suggest any kind of crude conspiracy, but to draw attention to the fact that an inventor's own cultural values become embedded in all technological objects. What is the use in seeing how we embed values and choices in technology? That is, in seeing that it isn't some outside force bearing down on us with its own logic.

If we don't see this embedding, we risk becoming passive and developing a dangerous apathy. It focuses our minds on how to adapt to technology, not on how to shape it. And so removes from our public discourse a vital aspect of how we live. This creates a pressing need for us as citizens to understand deeply how technology comes about. Not just simply to grasp the impressive world of technology, but to exercise the civic duty of shaping those forces that form our lives so intimately, deeply, and lastingly.

Electric Cars

I'M OFTEN ASKED "why don't we have electric cars?" The tone of the question usually implies some government-industrial conspiracy, led by the auto makers. I find them, though, blameless: General Motors and Toyota put a half billion dollars into electric cars. And they each brought out one that flopped. The real culprit is much less interesting. It's batteries.

Finding powerful enough batteries is the Achilles' heel of electric cars. You can visualize the problem like this: A typical car carries one hundred pounds of gas, which moves the car about 250 miles. To have the same range, an electric car would need 1,000 pounds of batteries— over four times as much. This means the whole car would be filled with batteries. The bad news is that there is no battery breakthrough on the horizon to fix this problem.

So, is the gasoline powered engine always to rule the roadways? Is there no hope of a takeover by electric cars? The good news is electric cars are succeeding, but only after taking a tip from that great political philosopher Machiavelli.

He advised ambitious Princes that "[t]he best and most certain way" to take over a dominion was to "take up residence there." And that's exactly what's happening: Hybrid cars that combine electric and gas powered motors are now appearing on the road.

These hybrids bring out the best of both engines. The weakness of the gas engine is the energy it wastes. Only about a quarter of the energy in the tank is used to move the car forward, most is lost during breaking, and as heat radiating from the car. Keep that in mind the next time you fill your tank: For every ten bucks of gas you put in, your engine tosses away seven dollars and fifty cents.

In a hybrid, though, when the brakes are hit, and energy wasted, the electric motor comes into play. During braking the gas engine uses its surplus energy to charge the batteries. Then when the hybrid car accelerates, the electric motor uses this power to assist the gas engine, allowing that engine to be smaller. This decreases the amount of gas used, so much so that some hybrid cars get 60 miles to the gallon. For most people this would mean one stop a month at the gas station.

Will these new hybrid cars succeed? To do so, they must hit consumers in the pocketbook—and these hybrids don't yet do it. Although current models save fuel and exhaust fewer pollutants than conventional cars, these savings still don't match the price difference between a hybrid and a conventional car. But keep your eyes open as every auto maker brings out a hybrid electric and gas car, and competes for market share.

Air Conditioning

HERE IN CENTRAL ILLINOIS, as we reach heat indices over 100, it would seem obvious that once invented air-conditioning would have quickly spread to homes and offices. Yet the answer is no. It had to battle a major trend of early 20TH century America: The open air movement. Many hygienists at the turn of the nineteenth century believed indoor air unhealthy because of respiration. They pictured large crowds of people spewing out toxic carbon dioxide. So, they opposed air-conditioning claiming it was something for a factory, not for the home.

Air-conditioning came from the textile industry. Its name comes from "yarn-conditioning". That's where textiles are exposed to moist air so they can be easily stretched. In the early 1900s textile mills installed high-speed looms that produced tremendous heat and dried the air, which could no longer soften the threads. These brittle threads easily broke. To prevent this, the textile engineers invented air-conditioning to control the humidity in the air.

Air conditioning then spread to other industries where humidity, control was important: In high humidity chewing gum wouldn't congeal, chocolate gets a gray powdery coating from the fat rising to the surface, and bread dough "slimes" and goes sour.

The open-air movement blocked the export of air-conditioning from the factory, especially to schools. At that time open air schools were popular for tubercular children, and were being extended to all students. In the winter the students wore coats, gloves and hats, and were wrapped to their seats with special padded bags for insulation. And there was no mechanical ventilation in summer. So, blocked in homes, office and schools how did air-conditioning take over our living spaces?

Via a growing cultural phenomenon: the Cinema. A few brave theater owners tried air-conditioning in the early 1930s and found it attracted patrons in droves. By 1940 ninety percent of all movie theaters had air-conditioning. And once patrons realized they weren't dying from it, air-conditioning spread to homes and offices.

Now, of course, it's essential to every aspect of American life. Without air-conditioning there would be no microchips, no Dallas Texas as a world financial center, and the National Hockey League's Phoenix Arizona Coyotes would still be Canada's Winnipeg Jets.

Hammond Organ

MY GRANDMOTHER HAD A NEARLY sacred object in her home. We were forbidden to touch it, except under supervision. It was her Hammond organ. We were, of course, allowed to listen to her play it. I can still picture her sitting at her Hammond, smelling of lilac face powder as she pounded out show tunes and pop favorites.

Now to her, this Hammond was an expensive treasure, but the real story of the Hammond organ is that it made this instrument available to nearly everyone.

Laurens Hammond, an electrical engineer invented the Hammond organ out of desperation. Hammond failed, in the 1920s, to market his invention for making 3-D movies. It was a motor-driven contraption that covered one eye, then the other. He needed to make something more practical. He turned to making electric clocks.

He based his clock on an electric motor he'd invented. It rotated at a very constant speed, exactly what is needed for a clock. Hammond did well at first, but soon others entered the market and he found himself losing money until by 1935 he was half a million dollars in debt.

In his desperation Hammond realized that his motor could be used to make sound. Hammond knew that sound from, say, a violin string arose because the string vibrates at a constant rate—moving at a constant rate was exactly what his motor did. Inspired by his boyhood memories of a church organ he made an electric organ.

Hammond inserted a long shaft into his special motor. He placed along it what he called "tone wheels". Each wheel had a different number of metal points on it. The motor spun these points past magnets that generated sound.

In 1935 Hammond debuted his organ at Rockefeller Center. George Gershwin, reportedly, bought one on the spot. Soon Hammond Organs were in Madison Square Garden and the Hollywood bowl, and then everywhere from skating rinks to funeral parlors.

The organ succeeded because of its low cost and convenience compared to a pipe organ. Unlike a pipe organ, a Hammond was never out of tune, was lightweight, and cost less than a tenth the price. By 1950 the Hammond was in homes across America, although critics complained about its sound. One characterized the Hammond's sound as "lifeless, dull, dead, hooty, [and] tubby."

Perhaps, but not in the hands of the artists who made the Hammond a true instrument. Jack McDuff, a true master, who could make the Hammond B3 Organ swing.

Cigarette Machine

IF I HAD TO NAME AN INVENTOR who'd made the largest impact on the last century, I'd consider James A. Bonsack. The economic impact of his work might put him up there with Thomas Edison, Alexander Graham Bell, and the flying Wright brothers.

In 1881 Bonsack invented the automatic cigarette making machine. It helped tobacco use skyrocket in the 20TH century.

Using an incredible system of gears, rollers, and levers, it made one hundred thousand cigarettes a day. As Bonsack noted in his patent application "This general result has heretofore been attempted, but so far as I know with but little success." Indeed, an efficient way to make cigarettes was the stumbling block to bringing tobacco to the masses.

In the years before Bonsack's invention, in the early 1870s, tobacco consumption had fallen to an all-time low, yet cigarette sales were outstripping demand. Growers had learned that nicotine delivered by inhalation is a highly addicting substance. Once inside the body nicotine is absorbed by the vast surface of the lungs, and passes rapidly into the bloodstream. From there is it carried back to the heart, which sends a large dose directly, and undiluted, to the brain. The brain, takes in all of the nicotine carried to it. This process take only seven seconds. Compare that to heroin, which, when injected in the forearm, takes 14 seconds.

So, growers worked out a way to cure the smoke so that it could be taken into the lungs, unlike cigar and pipe tobacco, which are too harsh. This new tobacco created a demand for cigarettes, the ideal vehicle for inhaling nicotine. James Bonsack's machine met the need with flying colors.

Perhaps it isn't fair to blame Bonsack entirely for this. The demand for cigarettes reflected the spirit of the age. A cigar and a pipe were smoked slowly and leisurely, but a cigarette reflected the quickening pace of our nascent industrial age. A cigarette was "light, quick, and short", a "potent symbol of the new velocity of modern life."

With Bonsack's speedy automatic cigarette making machine, manufacturers helped Americans to smoke over a billion cigarettes by 1889, increasing twenty years later to over ten billion. Today American's smoke nearly a trillion cigarettes per year. This, of course, has taken a tremendous human and economic toll.

No wonder some have dubbed the 20TH century the "Cigarette Century." Let's hope the 21ST century gets a better title.

[*The quotations are from* Tastes of Paradise *by Wolfgang Shivelbusch published by Pantheon, New York, 1992.*]

Bicycles

PLEASE WATCH OUT IF YOU live in my neighborhood. After twenty years of driving to work I've rediscovered my bicycle. Now I zoom into work in four minutes and thirty seconds, yelling the whole way at anyone who walks in my bike lane.

The bicycle, although invented 5,000 years after the wheel, was the first effective human-propelled vehicle. A human being on a bike is the most efficient way to move a weight around. A human just walking on the ground takes about as much energy, per pound, as a horse; but give that person a bike and they'll use one-fifth of the energy to move that same pound. On this basis it makes them more efficient than even a jet.

The secret is the bike's ingenious design. It takes maximum advantage of the strengths of the human body. A bike uses the most powerful muscles—the thighs—and it uses them in just the right motion. There is nothing more natural than the smooth rotary action of the feet. And on the bike they rotate at an ideal seventy revolutions per minute. The result: A single turn of the pedals advances the bike sixteen feet or so making it extremely easy to get from here to there.

The bike truly revolutionized our world. It accelerated the proliferation of machines in our lives. It did this by creating a huge demand for mass-produced precision parts, the average bike uses about 1,000 individual pieces. This brought forth new methods to produce metal parts. For example, bicycles needed bearings to make their wheels spin. This created a whole industry, so that by the turn of the 20TH century bearings were available for all sort of machines.

Also, the bicycle directed our minds toward the possibilities of independent long distance travel. Once this possibility appeared in our minds, the bike led conceptually to the car, and all sorts of other motorized vehicles.

But to me, none of these are more wonderful than the bike. It demands few energy resources, contributes little to pollution, and makes a positive contribution to health. It can be regarded as the most benevolent of machines—as long as you stay out of my bike lane.

Concrete

RECENTLY WHILE WAITING FOR AN appointment, I starred out the window and watched an ancient art: The making of concrete.

The origin of the word gives away its ancientness. It's made by combining the Latin prefix *com* meaning "together," and *crescere* meaning "to grow." The names comes about because when the ingredients making up concrete—water, gravel, sand, and a bit of cement—are mixed they turn into a hard, rigid solid.

The Romans discovered concrete by accident. A builder was making some mortar and he happened to be working near Mount Vesuvius, the famous volcano. He tossed in some volcanic ash and noticed that when his mixture dried it made a very hard substance.

From this serendipitous beginning the Romans fine tuned the recipe for concrete. They mixed horse hair to reduce the amount it shrank during hardening; and they also added blood, which made the stuff frost-resistant. Today we use plastics for horse hair and special chemicals instead of blood, but the same principles apply.

With these innovations Roman concrete reached a level of quality unmatched until this century. Just look at the Pantheon, the Temple of the Gods.

The Roman Emperor and builder Hadrian capped its rotunda with a concrete dome 144 feet in diameter. Michelangelo found it so beautiful he called it "angelic" and declared it "not of human design."

Although the Pantheon is the most visible example of Roman concrete engineering, they also made incredible structures underwater. An amazing property of concrete is that it can dry and harden underwater. So, the Romans used it to make piers, breakwaters, and lighthouse foundations that differ little from the ones we build today.

In our age the concrete industry is so important that it now takes up about 10 percent of our gross national product. And it promises to take up even more.

Perhaps the most bizarre application for concrete is in making ships and submarines. Although not well known concrete will float if you add enough air to it. A concrete submarine can dive deeper than a metal one because concrete is very strong under pressure. Once submerged it would be hard to detect. A concrete submarine fools sonar into thinking its the ocean floor.

Now, concrete is becoming high tech. Engineers are inventing smart concrete that can conduct electrical signals. It'll be able to detect vehicles, perhaps even guide one down a highway. And when used in buildings it might even detect earthquakes. Although I note that the Romans needed no such thing: After 2,000 years the concrete dome of the Pantheon is still standing.

Technological Optimism

WHILE BROWSING AT A USED BOOKSTORE a title caught my eye: *The Boy Engineer*. It's a 1950s book to entice boys to become engineers.

The book wears its era boldly: From the sexism of its title to its attitude toward engineers. A technological optimism fills every page: The book suggests a boy become an engineer to "solve problems of water shortages, traffic congestion, and to find cheaper power."

We certainly don't think like this today as we face some of the down sides to technology: bombs, high tech genocide, and perhaps environmental damage. This book, *The Boy Engineer*, published in 1959, marks the end of an era begun in the 1920s, when every new engineering marvel fascinated us. To see this listen to Ella Fitzgerald singing Cole Porter's hit *You're the Top*. She compares her lover to every great thing in the world of 1934:

"You're the top, you're Mahatma Ma Gandhi.

You're the top, you're Napoleon Brandy, you're the purple light of a summer night in Spain, you're the National Gallery, you're Crosby's salary, you're cellophane ...."

Cellophane? Sure Gandhi, he'd reached his worldwide fame only a few years before, and Bing Crosby was a major star, but Cellophane? Yes.

Jaded today by technology we find it hard to imagine mundane cellophane taking America by storm in the 1920s and 30s. In those decades, practically anything wrapped in cellophane sold better. The sales of cellophane-wrapped handkerchiefs rose nearly one hundred percent, marshmallows one thousand percent, and donuts sales jumped two thousand percent. Occasionally even people wrapped themselves in it: A magazine editor held a press conference shrouded in cellophane, and a Hollywood star paid ten thousand dollars for a black cellophane hat trimmed with diamonds. So pervasive was cellophane that in a *New Yorker* cartoon a nurse shows a new baby to its father who gasps "My word! No cellophane?"

I don't think we've felt such uncritical fascination with technology in years; we've gone through a time where to many the products of engineers are the chief source of any discontent. I think both attitudes are wrong-headed.

They miss that our sense of the positives and negatives of technology reflect ourselves. My reaction, as an engineer, is to marvel at the details of any technology that changes my life, but also to feel sadness in the loss of old ways. This reflects my own nature: I usually esteem what is familiar, and avoid large doses of the totally new. That's why I usually prefer used bookstores to new ones.

Technology & Terrorism

THE WORLD TRADE CENTER ATTACK REVEALS clearly the fragility of the incredible, nearly invisible, web of technology that surrounds us.

The attack severed the web of airplanes, antennas, and cables that connect our world. Packages stopped moving, people stayed put, and telephones were silenced. We've learned by its absence how this technological mesh makes our lives easier, but paradoxically we now see that it makes terrorism easier also.

At first blush the attacks seem very low tech. We expected a rogue nation to lob a crude nuclear bomb, instead we got a relic from World War II: A kamikaze driving a jet. But make no mistake about it, our high tech web enabled every aspect of the attack on the World Trade Centers. Their weapon, of course, was a technologically sophisticated jetliner, but the enabling goes even deeper. As an example, consider the central role of the computer in training the pilots.

When screaming through the sky at four hundred and fifty miles per hour the towers of the World Trade center looked to the pilot like a needle. Over a mile away he made exactly the right maneuvers to be aligned with his wafer-thin target. To learn to do this the terrorists took advantage of computer technology.

They employed the computer flight simulators used today to train commercial pilots. These machines duplicate the cockpit of a jetliner. So exact are these simulations that often a pilot's first flight in a real, live jet is with paying passengers. They allow a pilot to mimic taking off, flying and landing a jumbo jet. Pilots can practice navigating around urban areas because the simulators reproduce the topography of every major city in the world. But the simulators also let two terrorist pilots practice, again and again, crashing into the World

Trade Centers. All without the necessity of owning a jet.

The lesson from this is not the evil of technology. The attack reveals the truth of the aphorism "technology is neither good, nor bad; nor is it neutral." It depends on the human at the controls. The attack should caution us from looking for some quick technological fix to the problems of terrorism. We want some ultra-sensitive metal detector that will unmask any weapon, or some new x-ray machine that exposes all dangers. Yet, the only lasting solution will be a human one: At the basest level an alert and thinking person at an airport, and at its deepest, agreement and understanding among peoples.

Anthrax

A LTHOUGH ANTHRAX SEEMS A NEW, high tech weapon of terror, the history of biological weapons goes back to the beginning of time.

So repellent were these practices that the Romans forbade their use in their "Law of Nations." Yet, of course, the practice continued. As a 20TH century bio-warrior put it, "If it's important enough to be included in a treaty, it must be worth having in your arsenal."

In the 14TH century plague-ridden corpses of Tartar soldiers were catapulted over walls of besieged cities. In the 16TH century Germans burned shredded hooves and horns to make toxic clouds. And in the 18TH century, during the French and Indian Wars, a British commander gave smallpox infested blankets to native Americans.

The potency of biological weapons has increased in our century because of the astonishing growth of molecular biology. Consider anthrax.

In its natural form the disease usually infects sheep or cattle. It causes black lesions to develop, hence the name anthrax from the Greek for coal or carbon. What attracted the military to the anthrax bacterium was it hardiness. When not actively infecting an animal, it forms hard-shelled spores that survive in soil for at least eighty years. In fact, in World War II, if Britain had bombed Berlin with anthrax, instead of explosives, the city would likely still be infected today.

Using a great deal of technique we've figured out how to take this naturally occurring bacterium and, in military parlance, weaponize it.

To make weapons' grade anthrax takes a great deal of expertise. It's size must be carefully controlled by coating the particles. This coating makes the particles large enough to catch the wind and float, yet small enough to penetrate the natural defenses of our respiratory systems. It is this refined anthrax that is most terrifying: Once in this form it can be delivered by aerosol spray or crop-dusters.

While it may be tempting to blame technology for this terror, bear in mind that just as our molecular biology allows a highly advanced form of anthrax to be made, the same science has brought forth high-powered antibiotics and vaccines to combat the disease. As always the evil doesn't lie in the tool, it's where it always has been: inside human beings.

First E-mail Message

I BELIEVE THAT THE QUALITY OF COMMUNICATION has declined. Here's my evidence: Samuel Morse's first telegraph message in 1844 was the weighty question "What hath God wrought?" Then, in 1876, Alexander Graham Bell advanced communications, but degraded their quality with his first phone call. Nothing weighty at all, just "Mr. Watson, come here; I want you." Then in 1971 came the first e-mail message. Its author cannot quite remember it exactly, but he thinks it was "Q-W-E-R-T-Y-I-O-P." The only thing the sender can remember for sure, is that it was all upper case.

It was sent by Ray Tomlinson, the inventor of e-mail. He worked as an engineer developing something called the ARPANET, a precursor to the Internet. The ARPANET, was designed to connect computers across the nation. It started with fifteen computers placed in California, Utah and New England.

Ray Tomlinson, who worked on two of the computers stored in New England, was in the habit of leaving messages for his co-workers on the computer. A kind of note tacked to the computer, although stored inside, of course. It could only be read by someone sitting at that local computer. Tomlinson realized it would be useful to send messages to the engineers working at computers in other states.

So he wrote a program to send e-mail, then sent his first test message—that "Q-W-E-R-T-Y-I-O-P"—from a computer on his left to one on the right. That first e-mail message traveled all of about five feet.

He showed his new e-mail system to a colleague and said "Don't tell anyone. This isn't what we're supposed to be working on." But as word leaked out he decided to announce his new program himself. He sent out e-mail messages letting his colleagues across the nation know about his work. The first real use of e-mail was a kind of spam that announced its own existence. This message also introduced a symbol that now permeates our world.

Tomlinson explained in his message, that each user needed to distinguish messages intended only for their local computer from those headed out into the computer network. He told them to use an @ sign.

Soon e-mail became the most popular use for the ARPANET. Within two years 75% of its traffic was e-mail. Ray Tomlinson's e-mail spawned a revolution as great as Samuel Morse's telegraph and Alexander Graham Bell's telephone, but will he be remembered? He doesn't think so, but today more than 125 million people have e-mail addresses. All featuring Ray Tomlinson's @ sign.

Face Recognition Software

THERE IS AN AMAZING TECHNOLOGY now available to fight crime. It's a computer program that can recognize faces.

It identifies a face by measuring how its features are combined. It examines, for example, the shape of the triangle made by the eyes and nose, or how high a cheekbone is above a chin. In this way it can match a face photographed by, say, a video surveillance camera to one of millions stored in the computer's memory.

It would seem that to rid us of much crime, all we need to do is to use these computers in all sorts of places. What image could be more comforting than this: Automatic face-recognition computers continuously scan an airport crowd. When it finds a terrorist, it alerts the authorities who swoop down and arrest him.

But before installing this everywhere we need to place this new technology in perspective. It seems new and revolutionary, but it's just one more step down a perilous path we've been taking since World War II. We feared at the end of the war the world of George Orwell's *1984*. But it isn't Orwell's Big Brother Police Force and their in-your-face technology that menaces us. We don't often worry about a brutal police force that operates as it pleases, since World War II we've moved step-by-step toward a system were a police state need no longer be brutal, or openly inquisitorial, or even omnipresent in public consciousness. Police have instead moved in the direction of anticipating and forestalling crime. So, the trend is toward tracking every citizen throughout his or her life—geographically, commercially, and biologically.

This began soon after World War II with records of fingers prints, extensive paper dossiers on citizens, and then computer punch cards to sort through files. It evolved into the electronic databases and biological profiling we have today. These new face recognition computer programs are just a way to quietly add a page to an electronic dossier.

Still, the potential for abuse is enormous. In the future, perhaps, when someone approaches a sales desk their credit info would be displayed automatically for the sales staff. Or, with enough cameras, the state could track the public movements of everyone. As a result people would be less likely to do public activities, to engage, for example, in protests that offend powerful interests.

So, face recognition technologies may seem a non-intrusive, painless way to keep order. But as we decide how to use them we must keep in mind the story of police work since World War II: The most insidious technique is the one which makes itself felt the least, and which represents the least burden, yet lets every citizen be thoroughly known to the state.

Mauve

R ARELY DO WE THINK OF A COLOR AS BEING invented by a
person, yet in the case of the color mauve this might be a
reasonable way to think about it. It started when thirteen year old
William Henry Perkin wanted to take a chemistry class in London.

His father thought chemistry a waste of time, and even though the
class met only at lunch he refused to pay for it. Henry persisted until
his father gave in. This course sparked an intense interest in
chemistry, so much so that Perkin wished to pursue it at the Royal
College of London. His father still saw no future in it, but again
Perkin persisted until he persuaded his father. Perkin started College
at age 15, and by 17 had finished the basic courses. This earned him a
job as a lab assistant to a chemistry professor.

He was given the job of making quinine, a drug used to fight
malaria, something important to an empire as vast as the British one.
While mixing up his chemicals, Perkin accidentally made a dark
sludge. He was about to throw it away when he noticed it stained his
bench cloth. It dyed the cloth what he called a "strangely beautiful"
color. It was an elusive color that shifted from pale violet to a deep
purplish red. When he tried to wash it out he couldn't, nor did it fade
over time. Perkin had made, accidentally, the first successful synthetic
dye.

At the time this was quite an achievement because the palette of
dyes was very limited. They came only from nature—shellfish, insects,
vegetables, and plants—and this made the colors very expensive. For
example, Tyrian Purple, a favorite of Royalty, cost a great deal because
it required crushing thousands of small mollusks imported from the
Mediterranean sea. Yet, now Perkin had made cheaply in his lab a
purple color. He named it *mauve*, after a French flower of a similar

color.

The young Perkin formed a company and began manufacturing his mauve dye. It became immensely popular when Queen Victoria wore it to her daughter's wedding. Soon the color appeared in everything from stamps to sausages. This allowed Perkin to retire a very rich man at age 36. He devoted his retirement to developing artificial scents for perfume.

Perkin's achievement, though, is beyond just bringing color to the world. He showed that chemistry could be useful and profitable. He paved the way for drugs designed by humans. Aspirin, for example, was derived from a chemical dye. So, in many ways Perkin was responsible for the enormous advances we've seen in our lifetime in medicine, perfume, food, explosives, and photography.

Pop Rocks Candy

YESTERDAY, WHILE SHOPPING, I noticed the reappearance of Pop Rocks, the most exciting candy from my childhood. To explain a Pop Rock I'll quote from its patent:

"When placed in the mouth [the candy] produces an entertaining but short-lived popping sensation.... The tingling effect in the mouth is sensational but short."

Its inventor, William Mitchell, had tried originally to make a powdered soft drink. He envisioned a packet of powder, which, when mixed in water made a carbonated, bubbly drink like a coke. It didn't work well, but, when he looked at the solid powder he saw bubbles in it. It was trapped carbon dioxide gas, the stuff that makes soda fizzy. When he tossed the solid powder in his mouth, it melted releasing the bubbles with a loud pop.

Mitchell's employer, General Foods, marketed his Pop Rocks candy in the early 70s. They were a smash hit. They came, as I recall, in the flavors Tingling Grape, Zinging Cherry, and Orbiting Orange. General Foods sold some five hundred million packets at fifteen cents a piece, but then disaster struck.

The violent pop of the candy delighted children, but scared parents. The FDA set up a Pop Rocks hot line with messages to allay parental fears, but rumors persisted. They increased when a shipment blew open the doors of an overheated delivery truck. In temperatures over 85 degrees Pop Rocks can pop on their own. The body blow against Pop Rocks was a rumor that still circulates today.

In the early 70s Quaker Oats promoted Life Cereal with a commercial featuring a boy named Mikey. A rumor spread that poor Mikey consumed some Pop Rocks, washed them down with a chaser of soda pop and his little stomach exploded, unable to handle the pressure from the carbon dioxide. Although totally untrue, the rumor took on a life of its own.

The huge number of inquiries prompted General Foods to send letters to fifty thousand school principals explaining that pop rocks were safe, and General Foods ran full-page ads in major newspapers. But none of this quelled public fears. By the early 80s General Foods threw in the towel and stopped producing Pop Rocks candy.

Now, a Spanish company has bought the rights to make Pop Rocks and is marketing them again. If, like me, you're a little sheepish, to buy goofy children's candy I have an alternative for you.

Visit the Cafe Atlantico in Washington D.C. and order the warm mushroom *ceviche*. The chef, there, is trying to "change people's dining experiences"; she wants them entertained as well as fed. So she's incorporated into her mushroom dish Pop Rocks.

Numbers

THIS IS MY 100TH PIECE FOR THIS STATION. I'm not celebrating, though, because right now I'm rejecting numbers. They've taken over my life and I'm trying to avoid them.

I got my first hint that numbers would dominate my world when my grade school teacher screened *Donald Duck in Mathmagicland*. While not great cinema, Donald showed how math permeates music, sports, and the arts. I should have been more alarmed when he quacked on and on, because today I find that numbers have indeed taken over. They begin their assault first thing in the morning. A single number determines what I'll wear that day. In the summer I use the heat index, in the winter, the wind chill factor; from that point on my day fills with numbers.

The information revolution alone has quadrupled the numbers in my life. I have three telephone numbers—two at home, one at my office—a pager, a cell phone, and a fax. Adding to this overload is my wife's work phone, cell phone, and pager. And to access my university library's computer system, I've memorized a fifteen digit ID number.

I'm not the first to complain about numbers filling our minds. In 1963 the New York Times objected to the newly introduced Zip Code for mail. They wrote "Life is rendered intolerably complex by substituting number-memory for name association, which is not germane to the human psyche." Indeed. No wonder I nearly lost it when the Post Office added four digits to the zip code. But, I held on to my sanity until I got a second phone line.

To accommodate our increased internet usage, my wife and I decided to get a second line dedicated to our computers. The telephone installers arrive and quickly did their work, finishing the job by handing me my new phone number. I stared at it, feeling numb.

Dazed, I went to our old phone to call a friend. Just as I finished dialing, the new phone rang. I hung up, rushed to answer it, only to find no one on the other end. I returned to the old phone. Again, just as I finished dialing the new phone rang. Suddenly I realized: I'm calling my new phone number. It had lodged itself in my overloaded mind.

So, you can see that celebrating this, my 100TH piece, frightens me a bit. Although my fear of numbers may come from elsewhere. I've just celebrated one of those birthdays that ends in a zero, one which marks a new decade of life. Perhaps that is what makes me want to stop counting. I keep telling myself, ala Jack Benny, I am only 39.

Leonardo da Vinci

W E THINK OF TECHNOLOGICAL INNOVATIONS as new, but the truth is much closer to the wisdom of Ecclesiastes: "What has been will be again, what has been done will be done again; there is nothing new under the sun."

The prime example of this is Leonardo Da Vinci. Not as an artist, but rather Leonardo Da Vinci the engineer.

It's not well known, but he was so fascinated by technology that he neglected his artistic work. One patron who sought to commission a painting wrote in exasperation "he is so much distracted from painting by his mathematical experiments as to become intolerant of the brush."

I'm happy for this, because I get great pleasure from browsing through his notebooks. What I love most is looking for those inventions that show how far he was ahead of his time.

He designed everything from helicopters to parachutes. There is even a strong hint that he designed a bicycle some 400 years before the first one appeared. One drawing, in particular, has always fascinated me. It is a plan for a most remarkable bridge.

It came about in 1500, when Ambassadors from the Ottoman Empire came to Rome looking for Italian engineers. They needed to build a bridge over the Golden Horn, an inlet of the Bosporus Strait near Istanbul. Leonardo Da Vinci offered his services as an engineer.

For the Ottoman Ambassadors he designed a bridge of solid stone. Its span of over 1,000 feet would have made it the longest bridge at the time. But the Ottoman Sultan took a look at Leonardo's design and rejected it. He and his advisors felt it too radical, and were sure it could never hold much weight, let alone ever be built.

To my eye Leonardo's bridge design is surprisingly modern. It is made of three arches, which look like a series of archers bows pulled back in parallel. Like all Da Vinci designs it combines perfectly function and aesthetics. Although the Sultan of the Ottoman Empire didn't believe in this bridge design, I'm pleased that today you can walk across that bridge.

In 1996 a Norwegian artist, Vebjoern Sand came across this bridge design in a touring exhibit of Leonardo's engineering drawings. Sond was overwhelmed by the beauty of it. To him it was the Mona Lisa of bridges. He started a campaign to convince the Norwegian Highway Department that the bridge could be built. He succeeded. Now, Leonardo's wonderful bridge stands triumphantly in a small town about twenty miles south of Oslo. Part of the triumph is that Leonardo's bridge isn't just a structure. It's a symbol that joins together art and science, usefulness and beauty. In sum, it is the representation of a great renaissance mind.

Swimsuits

How much does it cost to design a swimsuit? In at least one case, the answer is about a million dollars, and about three years of work. I'm talking about the Fastskin swimsuit, developed by Speedo, and worn by most swimmers in the recent Olympics.

This Fastskin suit is revolutionary: It helps a swimmer to shave about three percent off their performance time. Now, this may not sound like much, but when we're talking about world class athletes it can determine whether he or she wins a medal. To give you an idea: Thirty-one of thirty-two Olympic finals involved less than a three percent difference between first and sixth place.

What's the secret to this new swimsuit? Nature. Although it's inventor, Fiona Fairhurst, started by looking at human-made things. Fairhurst, a former competitive swimmer, and now an expert in textiles, studied race cars to see how they reduced drag from the wind. Next, she took up SCUBA diving to understand buoyancy. Then came a critical day in 1997 when she visited a group of military designers.

"I spotted a stuffed penguin," she says, "and they told me how they were trying to develop a flak jacket using principles of penguins' feathers. So I started thinking about what moves fast generally."

This led her to Cheetahs, and then dolphins. Finally, an expert at London's Natural History Museum convinced her that Sharks should be her model.

He told her how sharks create turbulence in the water like humans. They are very fast, but not naturally hydrodynamic. It's their skin that minimizes drag and helps them swim efficiently. A shark's skin has tiny triangular projections that point backwards. They decrease drag and turbulence by making water spiral off the shark's body.

So, Fairhurst magnified the skin of a shark and asked manufacturers to copy it. They made her fabric with ridges with the same proportions as on a shark. To get the right material took three years, 450 different types of fabric and 10 prototype swimsuits, each costing $25,000. So advanced is this new swimsuit that Fiona Fairhurst has been given the nickname "Q", after the character in the James Bond films who develops all the gadgets.

Now that she's the Queen of High Tech Swimsuits, what next? Rumor has it, she's looking into swimsuits that are sprayed on.

Legos

WHAT HAS BEEN CALLED A "disaster of modern life," and is claimed to have lead to the "demise of engineering?" Sir Harry Korto, a Nobel Prize winner in chemistry, claims it's this: The disappearance of Erector sets, and their replacement by Legos.

You may be too young to remember Erector sets. They are a toy made of perforated metal strips that can be held together with nuts and bolts to make nearly any shape.

Sir Harry, like most scientists and engineers over forty, played with the toy as a child. It helped him, he claims win his Noble Prize for figuring out the structure of a large molecule. He says that Legos would not have given him this skill.

I refuse to adjudicate this dispute, except to note that I had both an erector set and Legos, which shows that either may led to no good whatsoever.

2001 is the 100TH anniversary of the Erector set. The first was built in Britain by Frank Hornby. He was inspired by the cranes in Liverpool's dockyards. But the real hero is a man named A.C. Gilbert. In 1909, Gilbert had just graduated from the Yale Medical School, but he didn't want to practice as a doctor. He wanted success in a field that had been a hobby since early childhood: magic. So, he created the Mysto Manufacturing Company to produce magic sets for kids. A few years later it became the A.C. Gilbert Company, and moved into manufacturing all sorts of scientific toys and kits for children.

He got the idea for the Erector set in 1913. He lived in Connecticut and often took the train to New York City. One day he saw workers positioning and riveting the steel beams of an electrical tower. This inspired him to design and market what he called the "Mysto Erector Structural Steel Build"—soon shortened to just "Erector set."

It was a huge hit: Kids would beg Santa for an upgrade, especially the number 12 ½ deluxe kit, which came with blueprints for the "Mysterious Walking Giant Robot."

He soon expanded into a glass blowing kit and a chemistry set. And, in 1950 he introduced the Gilbert Atomic Energy Lab. It came complete with radioactive material and an accurate Geiger Counter.

But soon after Gilbert debuted the Atomic lab, sales of all these sets slowly declined, including his best seller the Erector Set. After 50 million units were sold, plastic took over. In 1958, Lego started selling its colored blocks that, along with other plastic toys, began putting erector sets out of business.

There is good news, though. Erector sets are poised for a comeback. The Brio Corporation has purchased the rights to the Erector set and is producing them again. So, perhaps this holiday season, some child will get a set and it will be the first of many startling engineering feats.

The Beatles

GEORGE HARRISON'S DEATH takes away a musical voice, but the influence of the Beatles still permeates today's music. The Beatles revolutionized the technology of making music.

Usually the opposite argument is made: That technology alone changes music. For example, Bing Crosby's quiet style of crooning could not have succeeded without the microphone. But the Beatles' work shows that at times musicians themselves can drive changes in technology.

The Beatles changed the recording studio from a place where live performances were transcribed to a workshop. Until the 1950s everyone from the singer to the orchestra crammed into a studio. They crowded around a microphone as they recorded direct to disc.

By the 1950s tape recorders changed this: The orchestra could be taped separately from the singer and then combined later. None realized the power of this more than the Beatles and their producer George Martin.

He visualized the making of a record as painting a picture in sound. It's no coincidence his favorite painters were the impressionists. The most influential technique of Martin and the Beatles—and still heard in today's music—is double-tracking.

To double track, a musician records a song twice, and then the sound engineer plays both takes together. This gives a richer more pleasing sound than a single take. The Beatles used this double tracking extensively, you can hear it clearly in one of their greatest hits: *Lucy in the Sky with Diamonds*. The driving force for all this double-tracking was John Lennon's desire to do anything to change his voice. On some songs he'd have it played backwards, on others he'd pass it through a rotating loud speaker to give his voice a whirling quality. But the trick he liked best was double-tracking. This, though, was laborious for the musicians and the engineers, but this changed after a late night recording session in 1964. Ken Townsend, a sound engineer, had an idea while driving home. He realized he could artificially produce double tracking by feeding the singer's voice to the tape recorder twice. Townsend delayed, though, the second feed with an electronic circuit, so that it was slightly different than the first feed. This gave the effect of taping the song twice, but took, of course, only one take. It was used on all Beatles albums from 1964 on.

This technological breakthrough gave the Beatles their unique sound, but its legacy is its impact on music technology. Because other bands wanted to sound like the Beatles, equipment manufacturers designed and sold tape records that automatically including this double tracking and many other Beatles techniques—all sparked by John, Paul, Ringo and, of course, George Harrison's extraordinary recording work in the Abbey Road studios.

Lord of the Rings

J.R.R. TOLKIEN'S *Lord of the Rings* enchants because it let's us escape into another world called "Middle Earth." Yet, odd as this fantasy world is, it carries an important message for our world.

Tolkien placed at the center of his saga the question of how technology fits into our lives.

The story appears to be about the quest to destroy a ring with incredible powers, but hidden not far below the surface is a clear message about technology.

Throughout the *Lord of the Rings* Tolkien often characterizes evil as technology. For example, one of the major villains, a wizard called Saruman, lives in a place Tolkien calls "Isengard." Tolkien, who was an Oxford Professor of Anglo-Saxon, knew Isengard meant "iron yard", what we might call an industrial park. Inside that iron yard the evil Wizard Saruman spends his days building mills, chopping down forest, and blowing things up. He creates a system of tunnels and dams, and vents for poisonous gases and fires. Tolkien writes that "wheels and engines and explosions always delight" Saurman and his followers. The idea of machines appears again when he describes the evil Saruman as having "a mind of metal and wheels."

In contrast to this evil were the Hobbits. A simple, small people who have an agrarian economy. Tolkien once wrote to a friend

> *"I am in fact a Hobbit (in all but size). I like gardens, trees and unmechanized farmlands; I smoke a pipe, and [I] like good plain food (unrefrigerated)* ....

Tolkien lived a life as opposed as possible to technology. During his lifetime he rejected trains, television and refrigerated food. He did own a car, but sold it at the beginning of World War II. By that time Tolkien perceived the damage cars and their new roads were doing to the landscape. He came to think of the internal combustion engine as the greatest evil ever put upon this Earth.

His experiences with war colored his view of technological change. He served in the trenches during World War II and experienced technology as fighter planes, tanks, bombings, and flame-throwers. "By 1918," he once said, "all but one of my close friends were dead."

Small wonder he disliked the immense power behind technology. In many ways the great theme of the *Lord of the Rings* is that no one should have dominion over the world. *The Lord of the Rings* is an anti-quest, with its goal to destroy universal power forever. Herein lies Tolkien's message to us, what make his *Lord of the Rings* still ring true today. He refused to let the material world draw the boundaries of life, and though his small Hobbits he asserted the individual's right and responsibility to shape the decisions and structures that determine their life.

Glowing Hockey Puck

L AST NIGHT AT A DINNER PARTY someone leaned over and said to me "How exactly do they make the hockey puck glow?" I thought they were drunk, but I found out that indeed on TV there's a hockey puck that glows blue, until it's slapped to over seventy miles an hour when it turns red and develops a long red tail. Intrigued, I looked into this and found it's a real *tour de force* to make a hockey puck glow.

Rick Cavallaro, who headed the team, described the problem as this: We had "to track and highlight a frozen hockey puck traveling at times in excess of 100 mph after being walloped by angry 250-pound men with sticks."

In developing the puck, the team's first lesson was that the laboratory isn't the same as the real world. In their labs they crammed a puck full of electronics, then gave it to a real-world player. He found the puck too light. Rick and his team learned a puck even an ounce lighter can change the game completely.

So, they got official hockey pucks, cut them in two, carved out the middles and loaded their batteries and transmitters. Putting them back together was their next major problem.

They searched to find a glue of just the the right weight and adhesiveness. They felt it pretty important the pucks not split apart and spew batteries on a slap shot. To test the various glues they hit pucks with sledge hammers, squeezed then in vises, and then shot them out of cannons at 105 miles per hour.

Then the real world intervened again. They didn't realize that the pucks are stored on ice for several hours before being put into play. The problem is that batteries lose their charge at low temperatures. Since in a hockey game up to thirty pucks might be used, they couldn't just turn them all on because when frozen the batteries wouldn't last long.

They put in a sensor so that when the referee drops it the shock turns the puck on. It goes off if it isn't hit for 45 seconds—which, of course, never happens in a game—but if the puck goes into the stands it turns off, and the new puck put into play by the ref takes over.

During the game the puck's electronics sends back it's position every few seconds to twelve computers housed in a forty-foot semi. They process all the data then draw the puck on the TV screen.

This new glowing puck upset hard core hockey fans, who claimed they could already follow it. But since it helped attract new viewers, I have bad news for devoted sports fans. The team that brought us the glowing hockey puck is adapting their technology to other sports. They want, for example, NASCAR coverage to feature cars that glow.

Corks

THE EXISTENCE OF MORE THAN forty species of birds, and other wildlife is in peril. Their survival depends on whether or not the wine cork survives. And yes, by that, I mean the stoppers used in wine bottles.

Corks has been used since the early 1600s when the Benedictine monk Dom Pérignon first used it to seal his bottles of sparkling wine. Today plastic threatens to replace natural cork.

Natural wine corks are made from the bark of a type of oak tree found in the western Mediterranean, mostly in Portugal, Spain, and Algeria.

All of the cork harvested in the Mediterranean is sold to Portugal, where a handful of producers make half of the world's annual supply of wine corks—thirteen billion to be exact.

To make a cork the manufacturers strip the bark from the trees, season it for six months or so, then boil it to kill mold and insects. The corks dry for three weeks in a warehouse, then are sliced into strips from which corks are punched out and polished. But then the trouble starts.

Traditionally they bleach corks in chlorine to kill bacteria and to improved the cork's appearance. What can happen, though, is a special mold can grow from the leftover chlorine.

It is this mold that has given plastic an inroad. The mold makes trichloroanisole, or TCA for short, which can ruin a bottle of wine. It makes it taste like like a cellar, or damp cardboard. It doesn't take much: A single tablespoon would destroy all the annual wine production of the U.S. Plastic corks, of course, don't form this mold.

This battle of natural cork versus plastic has serious consequences. Eighty-five percent of the world's wine corks come from Portugal.

This accounts for three percent of their GDP. A huge number of natural corks are still used, only one bottle in twenty has a plastic stopper, but the trend is toward plastic.

The battle isn't only for the well being of Portugal's economy. The increasing use of plastic stoppers puts wildlife at risk. For centuries the cork woodlands in Spain and Portugal have provided shelter for many species of birds. The cork forests provide sturdy, tranquil nesting sites, while the grasslands are ideal hunting grounds. Some birds have adapted to nesting almost solely in cork trees.

Cork farmers carefully nurture and sustain their trees because it takes twenty-five years for the bark to be good enough to harvest, after that they can harvest only every nine years. But if natural wine corks are no longer economically viable, the cork trees are not likely to be replanted as they naturally die out. Leaving large sections of natural forest land prey to other economic uses that don't need trees.

So, to save an endangered bird, go to your local wine shop and buy a bottle with a real cork.

Olympic Flame

ALTHOUGH THE OLYMPIC FLAME is an ancient tradition—it began at the earliest Olympic Games in Greece, where a sacred flame burned in honor of Zeus—it is now a very media savvy tradition. Television cameras cover the Torch's journey as 11,500 people carry it 13,500 miles across the nation. This created a huge headache for the man who designed this year's Torch. He's Stan Shelton, a mechanical engineering professor from Georgia Tech. Shelton is an expert in combustion, although this is the trickiest problem he's ever dealt with.

The torch must stay lit in hail, rain, high winds, at temperatures as low as twenty below zero. It must blaze brightly when carried by runners, but also when moved by automobile, airplane, train, ship, dog sled, skier, horse-drawn sleigh, snowmobile, ice skaters, and even prairie schooner. Not only must it stay lit, it must also be safe.

In 1956 an Australian runner entered the Melbourne Olympic Stadium carrying a spectacular torch shooting out flames like a huge sparkler. It burned the runner so severely, he had to miss the opening ceremony.

Starting in 1972, at the Munich games, designers moved to an Olympic Torch that is a very sophisticated cigarette lighter. It's fuel is pressurized gas that flows through a small opening.

Stan Shelton, the designer of this year's Torch, found it wasn't as simple as just making a cigarette lighter. The flame for example had to look good on television, even better than it did in the past. The standard butane used in a lighter is a very light blue, not photogenic at all! So, Shelton changed to the gas used by welders; it emits a bright orange flame, and can be made to rise up to twenty inches in the air. Enough for any camera to see.

Beauty was also a criterion in making the torch. Says Shelton, "I have always felt a kinship with artists, and I believe this torch is a great symbol of the marriage of artistic and technical creativity."

This year's torch looks like a fiery icicle. The body is tapered with an antique silver finish and dark-shaded grooves that run from top to bottom. The outer shell is made from aluminum and plated to produce a high-polished chrome finish. The torch is topped with a glass crown, a copper cauldron sits inside from which the Olympic flame rises.

The glass represents winter and nature as well as ice and purity. The polished chrome stands for modern technology, the aged silver the heritage of the West. And the copper represent Utah's history.

In addition to meeting all the criteria of beauty, safety and protecting the flame, Shelton had to meet one other very important condition. 11,500 torches were made, and they had to be made cheaply because every one of the torch bearers had the right to buy the torch they carried.

The Ice Hotel

I SPENT A NIGHT SLEEPING in one of the most incredible
structures in the world: The Ice Hotel in the Arctic Circle of
Sweden. My stay began with a tour.

"We have this check-in tour," said the guide, "because this is not a
normal hotel. And to be able to survive the night and enjoy your stay
here we have to teach you how to live here. Tonight it's very cold, it's
minus thirty degrees and its actually a bit dangerously cold, so you
have to be aware."

With that warning I entered the Ice Hotel, passing through two
reindeer skin covered doors set in a huge ice wall. Inside the Ice Hotel
all sound dies in the three feet thick snow and ice walls. The only
sound is my own footsteps.

Just inside is the Great Hallway. A mist rises from its floor; it's roof
made of great arches thirty feet tall, held up by massive pillars carved
from sparkling, clear ice. The arches look like they've been
transplanted form a great Gothic cathedral, and then frosted with
translucent snow and ice.

I survived the night, then in the morning tracked down the
Architect of the Ice Hotel, a man named Ake Larsson. I asked him
what the magnificent arches were made from. "I call," he said, "this
material *snice*—the density is just between snow and ice." To make
arches they use snow cannons to blast this special snice mixture into
arch shaped molds. But what inspires him to create these arches made
of snice?

"Yes,", he told me, "I spend the summers around Europe to look at
old cathedrals." I asked him how he felt about his Ice Cathedral
disappearing at the end of every winter. "Happy. Because then I start
to draw a new one."

Yet still I wondered why anyone would create something so intricate and beautiful, yet short lived. I returned to the Ice Hotel to search for the answer. I found an artist working on a huge ice statue of a women.

He told me that he was creating "Lady of the river." Adding with a Swedish accent "Yeah! Ah, a very big lady, the mother of the river."

The statue is very detailed, and has obviously taken days to carve. Yet, it will be gone in a few months. So I ask why does he like working in ice. "Yeah, its perfect," he said, "nobody can buy it, the water takes the life back. Yeah, it's perfect for me."

He tells me that because the water takes the life back this sculpture destroys illusion that art last forever, even bronze or granite eventually decays.

Now I see why someone would design an Ice Hotel, instead of a bricks and mortar building. A permanent building may be used for centuries, but its beauty is often ignored. Constructing a building of ice makes us appreciate it; no one can buy it to preserve it forever, so we are forced to cherish it in the moment.

Gridlock

THE WORD GRIDLOCK SEEMS TO HAVE been with us forever, but it is only of recent origin. It appeared first in 1980, and it was coined by an engineer. His name is Sam Schwartz, and he says he "hates traffic."

He began his professional career as a taxi driver in New York City. So it was natural for him to choose transportation engineering when he went to college. After graduation, he planned to return to New York City and, in his own words, "save the subways." But the Subway Transit Authority wouldn't give him a job, so he joined New York's Traffic Department. It was there he coined the term "Gridlock."

In 1971 John Lindsay, the Mayor of New York, proposed an ambitious "Red Zone" to reduce traffic in midtown Manhattan. This zone would ban all cars. Schwartz spent hours studying this plan with his partner, an old-time traffic engineer named Roy Cottam. They considered what would happen if the Red Zone plan were approved. They concluded Manhattan's "grid" of streets would "lock-up" and all traffic would grind to a halt. Schwartz recalls that very naturally they just transposed the words and the term "gridlock" was born.

They kept this term to themselves until 1980 when New York suffered a transit strike. Schwartz and his colleagues found themselves media stars—and their term gridlock becoming a catchword.

As the transit strike burst onto the front page, "gridlock" grabbed the attention of William Safire, the language maven at the *New York Times*. He called Sam to get the details, and featured him in his "On Language" column. Sam tried to share credit with his colleague Roy Cottam, but in Sam's words, Roy refused, "he didn't want to be blamed for it." Maybe this was wise, because it earned Schwartz the moniker "Gridlock Sam."

He's put it to the best use in his career. For years he wrote a popular traffic column for the *Daily News*, and published a book of secret routes for avoiding traffic jams.

Although Scwhartz has had a distinguished career—in 1985 he earned a Public Service Award, and in 1988 was honored as Transportation Engineer of the Year—he will forever be known as Gridlock Sam.

Head Skis

HOWARD HEAD, AN AIRCRAFT ENGINEER, first tried skiing in 1947. He loved it, but didn't ski as well as he liked. He said later "I didn't blame any lack of athletic ability for my bad skiing, but rather blamed the skis." Since Head was skiing in 1947 he had a legitimate complaint.

The most important aspect of a ski is its shape: the front and rear tips are raised to keep them from getting caught, and the ski itself is curved and arched so the ski can turn easily. In 1947, when Head first went down the slope, the skis were made of wood, so they lost their shape quickly, leaving the skier with little control. Because of this Head felt a ski should be made of metal. "If wood were the best material," he reasoned, "they'd still be making airplanes out of wood."

So Head put his expertise as an aircraft engineer to work. He invested two hundred and fifty dollars in a band saw and spent his spare time designing skis. He used a wooden core, then covered it on top and bottom with aluminum. It took Head six months to make his first ski, partly because to set the glue used to bond the metal and wood, he had to boil the ski in oil. This first ski started him on a quest that lasted three years.

In 1951, he entered the metal ski business full time, starting his company with six thousand dollars of poker winnings. So good was Head's metal ski that the public soon dubbed it the cheater because even beginning skiers could carve through turns with little effort. It turned the sport from one for very skilled athletes to one able to be mastered by thousands. Yet the public was slow to accept Head's skis.

To market them he tried to get professional skiers to use them in competition. Still, few paid much attention until in 1961 an unknown on the Swiss ski team won an Olympic race on a pair of Head metal skis. With additional praises from racers worldwide, the Head Ski Company soon dominated the market. Skis, though, weren't the only way Head revolutionized sports.

In 1970, at age 60, Head sold his company and retired. He took up tennis, but just like his skiing in 1947 he was unhappy with his performance: His off-center hits frustrated him, and he hated the the way his racket twisted in his hands.

So, six years into his retirement Head redesigned the tennis racket. He created one called The Prince Classic, an over-sized racket with a large sweet spot that gave more control and power. Just like his skis Howard Head's new tennis racket became a standard for the industry. So important were both these achievements that Head's first metal skis and his Prince Classic racket are now permanently displayed at the Smithsonian.

Gore-Tex

I HAVE A FAVORITE PAIR OF SHOES. With them I've stomped in puddles all over the world. I like them because they keep my feet dry. They're lined with the most amazing material: Gore-Tex.

It's also used in running suits, hiking boots, mountaineering gear, fishing equipment and ski outfits. Gore-Tex is usually advertised as waterproof, but really its secret is that it's breathable. There are many waterproof things we could wear; for example, a piece of plastic would seal out water, but it would also seal in everything else, making us feel clammy and uncomfortable. In contrast, Gore-Tex has billions of microscopic pores, which allow body heat to escape, but are small enough that water can't get in.

Gore-tex came about because Bill Gore, a chemical engineer fell in love with Teflon. The stuff that's used to make nonstick pots and pans.

Here's a quote from Bill Gore, paying tribute, as only a chemist could, to Teflon: "From a chemical point of view," he said, "it is unique, there is really nothing like it. You have carbon combined with fluorine, which is the most electronegative element, and the bond is one of the strongest in chemistry . . . It's also impervious to other chemicals, impervious to ultraviolet radiation, [and] resistant to high and low temperatures. In a sense, it is the ultimate all-American material." I wasn't aware these were the criteria for an all American, but his sentiment did get me Gore-Tex lined shoes.

Thrilled with the properties of Teflon, Bill Gore felt, to quote him, that "if we could ever unfold these molecules, get them to stretch out straight, we'd have a tremendous new kind of material." In fact, that's all that waterproof Gore-Tex is: Stretched Teflon. But this is not as simple as it sounds, as Gore found out when he started developing

Gore-Tex in his basement, with his son Robert. They took a rod of Teflon, heated it, then carefully stretched it. It snapped, a bit like the mozzarella cheese on pizza. Day after day they failed in their home laboratory. Then one night, angry after repeated failures, Robert Gore grabbed a rod of Teflon and violently and quickly stretched it. It didn't snap. Then and there Gore-Tex was born.

The first field test was done by Bill Gore and his wife Genevieve. In the summer of 1970 they fixed an old tent with Gore-Tex patches, then took their annual camping trip to Wyoming. The first night it rained, but the Gore-Tex kept them dry. Based on the success of this trial, the Gores started making mountaineering clothes, a logical choice since they were accomplished mountaineers, having climbed to 16,000 feet in the Himalayas. By 1978 word of mouth has spread to other outdoor enthusiasts, and soon Gore-Tex became an essential tool for all outdoor sports.

It's now even moving into areas unseen by Bill Gore. The All-American Gore-tex is becoming all-Scottish. The breathability and waterproofness of Gore-Tex make it ideal as the bag of a bagpipe.

The Queen of Voice mail

TODAY THERE IS PERHAPS NOTHING more annoying that dialing the phone and hearing this: "For help at any time press * h. Please enter extension and pound sign."

That's the voice of Jean Barbe, queen of the voice mail menu. Although voice mail is annoying, that wasn't the intent of its prolific inventor, engineer Gordon Matthews, author of some thirty patents. He said, "We didn't design this technology to annoy people, but rather to make their lives easier." In fact, he claimed that every one of his inventions came about because, to quote him "something bothered me." In the case of voice mail, it was a dumpster full of paper slips.

In the late 1970s Matthews owned a small telecommunications company. One day while waiting for a client he tried to call his home office, but because it was in a different time zone there was no one in. While waiting he noticed a dumpster full of those "While You Were Out" slips. Fusing these two observations—the inability to reach his office and the discarded slips—it struck him that there must be a better and more private way for a business to communicate internally. Then and there, he claimed, the idea for voice mail was born.

Now at the time there were answering machines that used cassette tapes, but Matthews had something much grander in mind. He wanted to store thousands of calls in a computer, which could then route them all over the company. Matthews recalls, "we were stretching technology" to make the system work. The Personal computer, so ubiquitous today, was just appearing and computing power was still an expensive thing. He built his first voice mail system from 64 telephones, 114 computer processors, and four refrigerator-sized drives to store the voice messages.

Today, of course, it's a crucial tool for all corporations. By 1989 he sold his company for millions and retired, spending a great deal of time on the golf course.

His retirement gave him time to reflect on his invention. Here's his final verdict and some advice on voice mail: "I'm not really pleased," he said, "with some of the things I see voice mail being used for today." Adding "If I call someone and have to go through four or five steps to reach a person, only to reach his message machine, I won't do business with him. People hide behind it." He probably could have used his creative talents to improve voice mail, but apparently he was bothered more by slow golfers.

Among his last patents was a golf cart system to prevent slow golfers from plugging up a course. He called it the Automatic Marshal. If golfers took too long at a hole, a monitoring system automatically alerted the golf course Marshal via pager.

Batteries in the Frig

I WAS WALKING THROUGH THE HALLWAYS when a colleague motioned to me. In a very conspiratorial whisper she said "I need your help." Then added quickly "do you know what bizarre thing my husband does?" I looked away, not wishing to know, but she continued. "He stores, in the refrigerator, batteries!"

Now, I listened carefully, because one of the most alarming things to me about marriage was to find batteries taking over the refrigerator: D batteries guarding the luncheon meat, C cells surrounding the milk, and double AAs scattered throughout conducting, apparently, some kind of reconnaissance. My colleague asked me "Can you as an engineer shed any light on the need for putting batteries in the refrigerator?"

So, what does science say about batteries in the refrigerator?

First, why would it even seem sensible to do this? It's because a battery generates a current by a chemical reaction. When the chemicals exhaust themselves, the battery dies. This reaction is only supposed to take place when the battery's being used but—and here is the inroad for the refrigerator/battery enthusiast spouse—the chemical reaction does go on even when the batteries not being used. Over time the reaction will corrode the battery, covering the end with a brown film. So, how do you stop this reaction? The answer: Lower the temperature.

So, this means batteries stored in the refrigerator will, in theory, last longer. So far, all seems to be in favor of the refrigerator/battery enthusiasts spouse. But, the key question is, how much longer will they last?

Consumers Report magazine took exactly 432 double A, C and D batteries. They stored some in the refrigerator, and some at room temperatures. At the end of five years they found that indeed the refrigerated batteries had more charge, but not by much. The room temperature batteries still had 96 percent of the charge of the refrigerated ones. So, is this enough to merit filling a refrigerator with batteries?

I suppose rational spouses could disagree, but to me it seems the answer is "no." Particularly when you consider the inconvenience of having to wait for the battery to warm up. Also, as the batteries come up to room temperature water condenses on the them, which could destroy electronic equipment.

Should you present your refrigerator/battery enthusiast spouse with these cold, hard scientific facts? No, I suggest you instead follow the advice of a poet. Ogden Nash once wrote: "To keep your marriage brimming, with love in the wedding cup, whenever you're wrong, admit it; whenever you're right, shut up." So, even though science is on our side, we will forever have batteries in our refrigerators. I'd suggest that as non-refrigerator battery storing spouses, we, instead, form a support group.

Fracture Mechanics

EVERY TIME I RIDE IN A JET, I look out the window and watch the wing. When I see it's still there, I say a silence thanks to Constance Tipper. We take for granted that in our high tech world, things like airplane wings, car axles and buildings stay together, but it hasn't always been this way; that they now stay together is largely because of Constance Tipper. Tipper was born at end of the 19TH century and followed a career path unusual at the time for a woman. She earned degrees in the sciences, then settled in at Cambridge university to study something that seemed very esoteric. She wanted to know exactly how the arrangement of the atoms in a metal affected its strength and durability. In her lab day after day, she used a special microscope to examine the structure of the metal, then studied how it broke. She did this quietly for nearly thirty years, until she was called to aid her country in its battle with Germany.

The German U-boats were sinking British ships at a rapid rate. Ship builders responded by developing an innovative way to make metal ships. Instead of riveting the slabs of metal, which was time consuming, they simply welded the pieces together. Using this new method they were able to produce a 10,000 ton ship in just forty-two days. During the war the shipyard produced nearly five thousand of these vessels, called Liberty Ships.

Although at first a welcome aid to the war effort, these ships soon became a liability. As they carried crucial supplies across the North Atlantic on the icy Archangel run, the keel of the ship would suddenly crack, as if it had turned to glass. This crack would propagate around the hull until the ship broke in two and foundered at sea.

This called into question whether the rapid welding method should be used. Perhaps, they thought, they should return to the older, slower riveting method. It was at this point that Constance Tipper entered the picture.

The British government appointed a committee to investigate the cause, with Tipper as the technical expert. Tipper, who'd investigated the failure of metals for years, pointed out that the ships fell apart in icy conditions. She acquired pieces of the failed ships, then returned to her lab and showed that under these icy conditions the steel rapidly became brittle, and could then snap like a dry twig. Her work revealed to the ship makers that the fault lay not in the welding, but in the steel. She showed them how to test the steel to ensure the stability of the ships.

With this work Tipper opened up a field called fracture mechanics. Its still used by engineers today to develop wings that don't fall off and car axles that stay attached. So, today as the mechanical world around you doesn't fall apart, give thanks to Constance Tipper.

Mood Rings

I RECENTLY CAME ACROSS A PATENT that described something that had fascinated me in my childhood. The patent described "devices for responding to thermal patterns and converting them to visible patterns." It took me a minute to realize that this was a very clunky way to describe Mood Rings.

Yes, I'm talking about that 1970s relic that changed colors, supposedly in response to the wearer's mood. The ring had a large glass stone that turned a golden yellow if you were tense, blue if you were happy, and purple for moodiness.

It worked because the bottom of the stone was covered with something called a liquid crystal. The metal band of a Mood Ring conducted heat from the finger to that temperature sensitive liquid crystal, which changed color in response to temperature of the skin. What surprised me was that the technology used to create Mood Rings was intended for something much grander.

The idea of using temperature-sensitive liquid crystals to build devices occurred to James Fergason in the late 1950s. As a young engineer at Westinghouse, he came across the substance while working on another project. It intrigued him, and, in his words, he "burrowed in" and "learned as much as possible about liquid crystals" in order to use them "to solve problems."

For the next thirty plus years, all that Fergason saw was uses for these liquid crystals. He designed a liner for a baby's bottle that indicated its temperature, and he proposed a bathtub mat that revealed, at a glance, the temperature of a shower. His first marketed invention, though, was a forehead band to measure the temperature of a person.

His greatest hope was to use the technology to save lives by diagnosing disease. He'd learned that a thin layer of these liquid crystals painted on the body would display a dramatic swirl of colors reflecting the temperatures under the skin. The patterns could reveal circulatory effects of diseases like hemophilia, arthritis, and diabetes. And the colors could reveal certain tumors. He even developed a way to screen for breast cancer. His methods caught on in France and Spain, but never in the U.S. Eventually, though, they were replaced everywhere by other methods such as x-ray and ultrasound.

Although Fergason's liquid crystal work didn't have the medical results he wanted, he has had a huge impact on our world beyond the 47 million mood rings sold. Today he says "I like to just sit on a plane and count how many LCD —Liquid Crystal Display—products there are in the in-flight magazines." This takes him a while. His liquid crystals are now in alarm clocks, radios calculators, laptop computer screens, and even make up the airliner's movie screen.

Disc vs Disk

HOW YOU SPELL THE WORD DISK affects how you view laws about copying recordings of music and movies. Lying in the balance is billions of dollars, and perhaps the cultural legacy we'll pass on to our heirs.

Disk with a "k" came about in the mid-17TH century, modeled on words like *whisk*. Disc with a "c" arose a half-century later from the Latin discus. Most people used these two spellings interchangeably until late in the 19TH century when they began using disc with a "c" to refer to phonograph records. This usage still persists in *compact disc*, spelled with a "c". Then in the 1940s, when engineers needed a term to describe the data storage devices of their computers they choose to spell disk with a "k." We still see this in the spelling of the *hard disk* of our computers.

Today, though, the distinction between these two spellings is no longer meaningful. In the past if you had a disc with a "c", like a phonograph record, there was no way it could become a part of your hard disk, spelled with a "k." If you had a record you could make a tape of it, but it was never as good as the original, and any copies of it were even worse. Now, of course, the digital revolution has erased the difference between the two spellings of disk. A computer can make a copy identical to the original.

This, of course, has the entertainment industry terrified, especially when combined with the Internet, which provides unlimited distribution of these digital copies. Right now they're using software tricks to reduce copying—certain CDs now work only in players that can't make copies, but they know these software solutions are only temporary. Computer hackers always come up with ways to evade these measures. To counter this threat the industry is turning to

Congress. They want intellectual property laws that make all copying illegal. We should be alarmed by these efforts.

We run the risk that every embodiment of thought or imagination may be subjected to some kind of commercial control. For example, as books become electronic, readers may lose the rights they've had since Gutenberg's time. The publishers of an electronic book can specify whether you can read the book all at once, or only in parts. And they can decide whether you read it once or a hundred times.

So, the risk is this: The literary and intellectual canon of the coming century may be locked into a digital vault accessible only to a few. As Congress writes laws to regulate digital intellectual property and copying, I think they should keep in mind an aphorism adapted from T.S. Eliot. "Good poets borrow, great poets steal."

[What Eliot actually said was this:

> *"One of the surest of tests is the way in which a poet borrows. Immature poets imitate; mature poets steal; bad poets deface what they take, and good poets make it into something better, or at least something different. The good poet welds his theft into a whole of feeling which is unique, utterly different from that from which it was torn; the bad poet throws it into something which has no cohesion. A good poet will usually borrow from authors remote in time, or alien in language, or diverse in interest. Chapman borrowed from Seneca; Shakespeare and Webster from Montaigne.* (Selected Essays. *Third enlarged edition. London: Faber & Faber, 1951.)*

Much richer than my simple paraphrase!]

Surveillance Tags

A S A BOY, ARTHUR MINASY used to shoplift. He stole marbles, erasers, pencils and Spalding tennis balls. He recalls that he and his friends would "kind of drop them in our knickers." All this would be unremarkable if Minasy hadn't grown up to create a multi-billion dollar industry that helps prevent theft.

In the early 1960s, tired of working as an electrical engineer for someone else, Minasy decided to start his own business. In his own words he "looked around to see what kind of industry I could possibly get into where there was something clever to be done." He noted that devices used by police to fight crime were very low tech. So he thought "that would be the place for me to carve a little niche."

The first thing he invented was something called the Vaicom, that's short for "variable image compositor." Its a machine that uses slides and a mirror so a witness can create, without an artist, a person's likeness.

But Minasy found selling to police departments a slow business. There was just too much bureaucracy. He felt he'd have an easier selling job if he turned to those most acquainted with thieves. So, he went to where the the thieves hung out: Stores.

To reach this market he invented the surveillance tags we still see today. Others had tried before him by loading tags, for example, with small amounts of radioactive material, which would be detected by a Geiger counter at the store's exit. But, reasonably, the government prohibited such a use of radioactivity. Minasy succeeded because he used a very simple approach. He crafted a plastic tag embedded with a metal coil, which was really a small antenna that responded to radio frequencies. Then he built a portal to surround a doorway which set off an alarm.

With one hundred thousand dollars of his own money he built a demonstration model into a metal brief case. The briefcase contained a miniature portal, through which he'd pass a surveillance tag which set off an irritating ringing sound. A message flashed on the lid of the briefcase: "Stop! Have inventory control tag removed at cashier's desk." He made his first sale to a Manhattan department store, doing nine thousand dollars worth of business in the first year. Soon enough sales of the device took off, spawning a three billion dollar industry.

Although its a sad commentary on our society, surveillance tags are now used all over the world: They've been attached to everything from clothing to parmesan cheese to hunks of salami to negotiable bonds on Wall Street. So completely did Minasy's surveillance tag become part of our culture that in 1991 when it was officially accepted into the permanent collection of the Smithsonian.

Slinky

THERE IS A SONG, A JINGLE, that's recognized by ninety percent of American adults. Its this: "It's Slinky, it's slinky, oh what a wonderful toy. It's Slinky, it's Slinky, fun for a girl and boy." Of course, that's the song for the slinky toy—a coil spring that walks down steps.

In 1943, Richard James, a naval engineer, was developing a spring that could keep sensitive instruments aboard warships steady in rough seas. Accidentally, he knocked one of his test springs off a shelf. It crawled, coil by coil, to a lower shelf, then onto a pile of books and finally came to rest on a table. This so enchanted James that when he got home he said to his wife "I think, if I could get the tension right, I could make it walk."

For two years Richard James worked to find the proper length and tension so it could walk perfectly down stairs. His wife, Betty, helped name this new spring. Flipping through the directory, she came across a word that meant "stealthy, sleek, and sinuous." That was, of course, "slinky."

By 1946, James had his Slinky ready to sell. On a snowy day, he set off for the Gimbels department store in Philadelphia to sell them. His wife worried the no one would want a Slinky. She gave a friend a dollar to buy the first one, so her husband wouldn't feel bad. When Richard and Betty James stepped off the elevator that day, they saw across the sales floor a mass of people waving $1 dollar bills. Within ninety minutes they'd sold 400 Slinkys. For the next fifteen years, Slinky continued to sell well. But Betty James saw little of the profit. Richard had, in her words, joined a "religious cult" and was giving them all the profits. By 1960 he left his wife and family to join this group in Bolivia, never to return to see his family again. Betty took over the Slinky company, now nearly bankrupt, and turned it into a multi-million dollar enterprise.

The company uses the same machines that Betty's husband designed in 1945. No one has ever been able to build better machines. So important are they to making Slinkys, that no one is ever allowed to photograph them, for fear foreign competitors could copy the machines and bootleg Slinkys.

The machines make a slinky in about ten seconds by coiling a 63 foot metal wire into 89 coils. When finished the machine drops the slinky which pops up, walks down a step, and steps into its own box. This isn't just for fun. A slinky wound too tight or too loose won't walk down the step and into the box correctly, and are rejected.

Two hundred and fifty million Slinkys, though, have walked into their own boxes since 1945. And Slinky shows no signs of slowing down: Sales have risen 25% in the last several years. So, it's likely a whole new generation will remember that Slinky Jingle.

Strobe Light Photography

I KNOW OF NO ENGINEER WHO'S changed the way we see the world as much as Harold Edgerton. It's unlikely you recognize his name, but because of him we carry in our minds an image of a milk drop splashing on a plate, or of a bullet piercing an object.

Edgerton invented the electronic flash used in photography. He's best known for using this flash to pioneer strobe photography to freeze time, and capture motions that occur in a fraction of second.

He started out by wanting to photograph only one thing: A motor. While doing research in 1931 at the Massachusetts Institute of Technology, he needed to see how a motor flexed as it rotated. Of course, a regular camera showed only a blur because in the time it took to open and close the shutter the motor had rotated hundreds of times. So, instead Edgerton darkened the room, opened the shutter permanently, and then exposed the film quickly with his flash.

In that moment, his career changed. The crystal clear, sharp photographs captured his imagination. He said once, "It shows you what's going on. There's no argument when you get through. No theory. It's the real world." For Edgerton, the strobe made the world intelligible, even simple. No longer did he study motors, but instead turned his camera toward the world. For the next sixty years he photographed everything he could, showing people things they'd never seen before.

He refined his flash unit until he could capture something happening in a millionth of a second. He froze the moment when a bat dented a baseball. He captured for the first time in mid-flight the wings of a hummingbird. And in perhaps his most famous photo he caught a drop of milk splashing off a table.

Although these images are indelibly impressed in our minds, Edgerton's strobe photography was serious scientific work. He once settled a lawsuit between two soap makers. One claimed the other had stolen their method for making soap power. Edgerton showed that while the two methods looked alike in motion, when frozen in time they were very different. In World War II he developed a special flash system to photograph the Normandy coast immediately before D-DAY, giving vital reconnaissance to U.S. forces. And he developed a method to capture how the mushroom cloud of an atomic bomb expanded and collapsed. Then he worked with Jacques Cousteau to build a flash system for underwater photography, using it to locate undersea wrecks.

Today, of course, Edgerton is best remembered as an artist. There is no serious collection of photographs that doesn't contain one of his strobe photos. Yet, however anonymous, his influence as an engineer is still felt. From the flash on your camera, to the flash that you see in a photocopier.

Thomas Stockham

IN EARLY JANUARY OF 2004 Thomas Stockham died. No doubt the name is completely unknown to you, yet he had a huge impact on our society. Due to Thomas Stockham we now have a generation that doesn't know what a vinyl record or a turntable is. He pioneered the science of digitally recording music, which is used to make compact discs.

When he first considered doing this in 1962, computers were too slow. At the time, it took nineteen minutes of computer processing to record a single second of music. But Stockham persisted. As he waited for computers to catch up with him, he worked out exactly what the computers should do with the digitally recorded music.

He studied something called "signal processing." To an electrical engineer, a signal is any kind of stimuli to our senses—light from a photograph, or sound from a record. Stockham focused on ways to improve these signals—a photograph can be blurry, or a sound might have extra noise like a hiss. An engineer calls all of these interferences noise. Stockham developed mathematical ways to remove this noise— to make a photograph clearer, or a sound more distinct.

He returned to recording music when a friend asked if Stockham could restore some antique Caruso records. Stockham jumped at the chance.

He used his methods to record the old 78's digitally, then removed all the pops, cracks and hisses. By 1972, he was able to use his techniques to examine the infamous eighteen and a half minute gap in the Watergate tapes. By this time, computers had become fast enough that he could record live music.

His big break came when he recorded the Cleveland Symphony Orchestra. After the session, the orchestra members crowded into a tiny playback room to judge the results. The technician hit the playback button and the room filled with a sound almost unnaturally clear and sharp. The musicians could hear the squeak of the piano pedals and the turning of sheet music. As one musician said, "It was like being on the stage."

As Stockham recalled after this session, digital recording "became afire then; people who had never talked to me before started calling me on the phone and saying, why didn't you tell us it could be this good?" Over the next few years he recorded over five hundred records. Today, of course, digital recording and CDs have taken over the industry.

Stockham's methods have touched all aspects of our world. His mathematical methods are used in playing DVDs, processing Hubble space telescope images, filtering spy photos, improving medical images, and creating better hearing aids. But of course, it's in the entertainment industry that he had the most impact. Small wonder he's the only electrical engineering professor to have won a Grammy, an Emmy, and an Academy Award.

Terry Bicycles

RECENTLY MY WIFE BOUGHT a bicycle designed especially for a woman. By this I don't mean it was missing a crossbar, but that the whole bike was redesigned with a women's anatomy in mind.

Intrigued by this, I called the bike's designer to find out how she got into designing bikes. I learned from, Georgena Terry, the bike's designer, that the bikes came about because of her restlessness and her love for blowtorches.

Terry began her career with a degree in theater, working behind the scenes, building sets and adjusting the lighting. But this didn't fulfill her, so she when to Wharton, earned an M.B.A. and worked as a stockbroker. But, as she told me, she hated "being tied down to a desk", it made her "absolutely nuts."

So, she quit her job, and returned to school to become an engineer. She found it "comforting to work with engineering and science because they tell the truth" because, as she told me, "they are very, very logical." I suppose, yet the next thing she did defies logic: She fell in love with a special type of welding called brazing. And here, of course, is where the blowtorch enters.

For her senior project she had to weld a bicycle frame from metal tubes. This was so intriguing to her that two years after graduation, she quit her well-paying engineering job and starting making bike frames for a living. As she said to me "sometimes I don't think a lot, I just go on a hunch, and I find that that's not always bad."

When she brought her bikes to rallies, women would approach her and ask, "can you build a bicycle for me." They pointed out that their bikes gave them neck and shoulder pain. She'd never thought of designing a bike specifically for women, but this intrigued her.

So, she went to a library to learn about the anatomical differences between men and women. She found out that the Wright-Patterson Air Force base had extensive data on these physical differences. The Air Force needed to design things like uniforms and cockpits to fit both genders.

She learned that a women isn't simply a smaller version of a man. For example, a women's upper body is proportionally longer than a man's. So, a bike that fits men in the legs and upper body, will fit women in only one of these areas. The key to making a women's bike, she decided, is getting them into a slightly more upright position.

In her first year, 1985, she sold twenty of these women's bikes, the following year, 1,300, then 5,000, and today its a multi-million dollar enterprise.

Now that her company is thriving, I asked what she does everyday. She told me "anything I want", noting that she since 1988 she hasn't had to use a blowtorch.

J.R. Pierce & Satellites

To me, to think of a satellite orbiting the earth and beaming radio signals around the globe seems like something out of science fiction -- and in a sense it actually is.

The prime mover behind satellites was an engineer named John Robinson Pierce, who became interested in space as a teenager. He devoured the articles in *Amazing Stories*, the first science fiction magazine. As he said "My reading took me into a world in which spaceships and communication through space by radio were commonplace."

He became so involved in the world of science fiction that he even wrote a few stories himself. He didn't, though, lose, his grasp on the real world. He used the money he'd earned from writing stories to pay tuition for engineering school at CalTech.

From CalTech he moved to Bell Telephone Labs in New Jersey. At the time they were setting up a system using radio waves to replace telephone wires. This required the Lab to place antennas on hilltops every thirty miles or so because any hills would cut the signals off. Pierce was struck by the idea that it would be easier to communicate between the moon and the earth, than across the United States, adding "if only one could put the apparatus in place [on the moon]."

He though little more about this until an engineers club invited him to speak because of his fame as a science fiction writer. Their only restriction was that he must talk about space. So, he returned to his idea about communicating with the moon. Not satisfied to just make up fictions about it, he put his engineering training to work and made calculations to estimate the cost and the type of equipment needed. He noted the only thing missing: How to get the stuff in space. Then, in 1957, the Soviet Union sent up Sputnik, and the rocket age began.

Seeing now that space satellites were feasible, Pierce set out to prove his idea. A year after Sputnik he saw a photo of a military weather balloon. Its shiny surface, he felt, was perfect for a communications satellite. With a bit of pleading, he got permission to use the weather balloon.

Called the *Echo* system, it was launched on August 12, 1960. The first message to be broadcast was by President Eisenhower and was sent from California to New Jersey. This message bounced off the shiny surface of the balloon. Thus proving that satellite communication was possible.

And now it has revolutionized our world. Today, people think nothing of telephone calls or live telecasts from anywhere on Earth, but until Pierce's pioneering work, the former was difficult and the later unthinkable.

Nautilus Machines

I OCCASIONALLY GO THE GYM and use the exercise machines. I regard these machines as torture devices. So, I'm not surprised that their inventor has been described as "among the darkest and crankiest people you're every likely to met."

His name is Arthur Jones and he invented, over a period of twenty years, the Nautilus machine. It started in 1948 when he was working out at the YMCA in Tulsa, Oklahoma. While exercising with barbells he noticed how awkward his movements were, and how slowly he built muscle.

He noted that with barbells the force is always directed downward by gravity. So that in a curl for example, the weight became harder to lift as his arm reached the horizontal, then easier as he curled his arm until it was vertical. He concluded that there must be a way to keep the force constant and thus exercise his muscles completely. So, he set out to design an exercise machine, which he called "an attempt to create a thinking person's barbell." His goal was to exercise all of the muscles in his body. As he said of himself after years of working with barbells: "I had the arms and legs of a gorilla and the body of a spider monkey." Gorillas and monkeys were something he knew well. Over the twenty years he took to invent his exercise machine, he operated an airline that captured wild animals from South America and African and sold them to zoos and circuses. His machine came to fruition, when, in 1968, political turmoil in Africa prevented any further flights by him.

Now with free time on his hands, he built what he called "the blue monster", the first Nautilus machine. It applied his ideas of variable resistance to muscles by carefully designed spiral pulleys. They resembled a section of the chambered nautilus shell, and so Jones named his machine Nautilus.

The machine wasn't well received at first. He appeared to most as a crackpot: A middle-aged man from Oklahoma who said he'd built a machine that would revolutionize the exercise industry. As the experts scoffed, Jones took his machine to local high school weightlifting teams. They won state championships. Then he trained a bodybuilder who won the nation's top titles. Then in 1970 the Kansas City Chiefs ordered several, and the Nautilus took off: By the 1980s, annual sales were about $400 million.

He's now sold off his company, and lives on a 500 acre farm in Florida with some 2000 crocodiles, a few hundred snakes, and a gorilla. And he likes to joke that now that he's rich, he's no longer called a crackpot, just eccentric. Indeed.

Vaseline

R IGHT IN YOUR MEDICINE CABINET is one of the very first
products made from oil. It's that jar of vaseline, known
generically as petroleum jelly.

It began in the late 19TH century with a young chemist, Robert
Chesebrough, who sold kerosene. Chesebrough got his kerosene from
the oil of sperm whales, but by 1859 he was put out of business when,
in Titusville, Pennsylvania, the first successful oil well was installed.
This new "oil" attracted him to Titusville. While visiting he learned of
a gooey substance, called rod wax, that stuck to the drilling rigs. It
built up on drill bits as they punctured the ground. This rod wax was a
nuisance to the riggers. They just scrapped it off and tossed it away.
But rumors floated around that it had miraculous healing powers. So,
Chesebrough got a bucket, loaded it up with this black wax and took a
sample back to his Brooklyn lab.

As a chemist it didn't take him long to extract the key ingredient—
the translucent material we know as petroleum jelly. Today we
wouldn't get very excited about something like petroleum jelly, but in
Chesebrough's time, the only oils available where lard, goose grease or
garlic oil -- all of which spoiled and smelled awful. So Chesebrough's
nearly colorless, unspoilable, odorless oil seemed like a miracle.

To test it's miracle properties he inflicted cuts and burns on himself,
then covered them with his new gel. It did help, although at the time
no one realized that this was because it sealed out bacteria, thus
preventing infections.

Satisfied that his new grease had healing properties, he took to the
road with his own medicine show, but first he named his gel. Using
the German word for *water* (wasser) and the Greek word for *oil*
(elaion) he came up with "Vaseline." Next, he traveled around New

York State demonstrating his miracle vaseline. Before a rapt audience he'd burn his skin with acid or an open flame, then spread the clear jelly on his injuries, showing at the same time his past injuries, healed, he claimed, by his miracle product.

Amazingly this worked. Soon he was selling a jar a minute, and when these ran out people begged their druggists to order more from Chesebrough. It was used for everything: chest colds, chapped hands, nasal congestion, and even to remove stains from furniture. By the turn of the twenty century it had penetrated the American market and was entering Europe, making Chesebrough a very rich man.

Although its miracle properties were eventually debunked, Chesebrough himself was always a true believer. He lived to age 96, revealing shortly before he died the secret to his longevity: Every day of his life he ate a spoonful of vaseline.

Directional Sound

ONE WOMEN BELIEVES THAT A special sound she's developed may save thousands of lives every year. Her name is Deborah Withington, and she's a scientist at Leeds University in England.

She's developed a special sound that, when heard can be located quickly and accurately by the listener.

Her Eureka moment came when she was sitting in her car and heard a fire engine's siren. "I realized," she said, "I had not the faintest idea where it was coming from. Because of that, I was still in danger." Most people would let this drop, but Withington is a neuroscientist who specializes in how people hear sounds.

She first did some research to see if others also couldn't locate the sound. Her studies showed that more than half the time, motorists couldn't tell whether an approaching siren was directly behind or in front of them. This percentage is worse than if people were to just guess.

To solve this problem, she used her understanding of the brain to invent a sound that lets humans locate it easily.

It works by using a rich mixture of pitches. For a human to identify the direction of sound, a lot of information is required. When a sound comes from, say, the right side, it arrives first at the right ear, and is louder than in the left ear. But for us to perceive this difference in loudness we need a sound with lots of different pitches. For example, a rushing river or a waterfall. With this wide range of pitches, our brains have the maximum number of cues available to locate the sound.

Withington tested it out in a siren, and while it worked so that people could locate it, that sound just didn't have the authority of the old one, so she combined it was the whirl of a traditional siren and it worked perfectly.

And there are other uses for it. For example, she claims that people will look toward the sound involuntarily. So, the Munich airport is testing it in their fire alarms, hoping that people will be attracted to the sound of the alarm and thus able to navigate through smoke.

And in its most novel use, Withington hopes it can be used on security cameras. When a bank is robbed, she envisions the security camera emitting her new sound. This would force the thief to look straight at the camera, giving the police a good look a the culprit.

Still, to me the pinnacle of western technology would be to add this sound to my phone. I'm always setting it down somewhere, and to find it again I have to listen to it buzz and slowly home in on it. It would be nice to be directionally drawn toward it.

Engineering Heritage

YESTERDAY I LEARNED THAT I'M a member of the world's oldest profession. I read this in an article on "engineering" in the the Encyclopedia Britannica. Its main point is captured by this favorite quote of mine from a scientist. He said, "a good scientist is a person with original ideas. A good engineer is a person who makes a design that works with as few original ideas as possible." This is the essence of all good engineering today. In fact, this way of thinking is part of our western heritage.

We often hear of the Western canon in Literature, but we have another western heritage: engineering. When we trace our roots in literature we start with Homer's *Iliad* or *Odyssey*. But as an engineer I don't think of ancient Greece as a nation of poets, but as a nation of merchants who built wealth to hire engineers to build huge things.

We still see Greek architecture in our museums, banks and churches. But in many ways the engineering achievements of the Greeks pale when compared to the Romans, who were most proud of their work, especially the aqueducts. Here is the cry of the aqueducts Roman Administrator: "I ask you, just compare with the vast monuments of this vital aqueduct those useless pyramids, or the good-for-nothing tourists attractions of the Greeks!" Without realizing it, he was highlighting the great contribution of the Romans to engineering: They set the profession firmly on economic principles. And with this, they paved the way for our consumer culture.

Consider the aqueducts. The Roman people didn't really need them. For washing, drinking and irrigation they could get water from nearby. But the Roman people wanted huge baths: They were a way to escape the blazing Mediterranean summer, or get warm in winter—and a place to meet and converse. We'd compare the baths today to a coffee house or a sidewalk cafe. And they were truly for everyone. To quote the great historian Gibbon, for the cost of a small copper coin any Roman "could purchase the daily enjoyment of a scene of pomp and luxury which might excite the envy of the Kings of Asia."

To meet consumer demands for public baths, the engineers built elaborate aqueducts to bring, from miles away, some 220 million gallons of water a day. So, with the baths the Roman engineers began the trend toward a consumer culture based on luxury because the public baths weren't needed anymore than a soft drink or a microwave oven is today.

So, tonight take a few moments to pay homage to our western heritage from the Romans: Flip on your TV, click to a shopping channel and watch them hawk useless trinkets.

Cruise Control

SUCCESSFUL ENGINEERS THINK in images. Their minds occupy a nonverbal world, not easily reducible to words. It is this kind of thinking that Ralph Teetor, the inventor of cruise control, had in spades. What makes his story remarkable is that he was blind.

After losing his sight at age five, he developed an exceptional ability to visualize objects and distances. For example, Ralph had helped build and install the basketball hoops at his high school and this was enough for him to be able to amaze his friends by sinking basket after basket. And at his father's shop, Teetor learned to create things perfectly out of wood or metal. So remarkable was his prowess with tools that by age ten his father built him a workshop, outfitted with grinders, lathes, and drills.

After high school, Teetor decided to become an engineer. But because 1906 was not as enlightened as today, many colleges refused to even consider his application. Teetor had a cousin attending the University of Pennsylvania. So, Ralph visited the school and persuaded the Dean of Engineering to admit him. Teetor excelled in the mechanical engineering program and on graduation became an inventor.

In 1921, he invented the automatic transmission, although he was too far ahead of his time; it only appeared years later when his patent had expired. Teetor invented a new fishing rod and reel because his wrist grew tired from fly fishing; he also designed and patented new locks. All, of course, without a drawing of any sort—just images in his mind. The one invention that he timed perfectly was cruise control. During World War II, the government set a speed limit of 35 miles per hour to conserve gas and rubber tires. Most motorists found it difficult and boring to travel long distances at such a slow speed. He claimed, though, that he'd never have worked on it except for his patent attorney. The attorney's jerky driving made Teetor car sick.

The first cruise controls were elaborate mechanical devices that required every bit of Ralph's ability to visual objects and distances. By 1960 every major auto manufacturer had added this feature to their cars, and with the Energy Crisis of the early 1970s, cruise control became a permanent fixture.

What was the secret to Teetor's success? One of the engineers who worked with him on the first cruise control device asked "With all that you have been able to accomplish, what more do you think you would have done if you had been able to see?" Ralph replied, with a smile, "I probably couldn't have done as much, because I can concentrate, and you can't."

Geiger Counters

TODAY GEIGER COUNTERS SELL SO briskly that manufacturers can't keep up with demand. They're popular with people worried about nuclear terrorist attacks. But when Hans Geiger invented it, it was just a way to make his life easier.

Geiger worked in 1907 with the most prominent scientist of the time: Lord Ernest Rutherford. Geiger and Rutherford wanted to measure the number of subatomic particles, called alpha particles, emitted by radioactive substances like radium. Amazingly they did this by counting the particles by eye. Alpha particles are too small to see, so Geiger created a special screen that magnified, in a sense, an alpha particle: It flashed when a particle hit it. This was an exhausting business that had to be done in the dark—Geiger recalls his lab as a "gloomy cellar", where he sat for long periods with his eyes glued to a microscope. And his concentration was often broken by Lord Rutherford wandering through the lab singing *Onward Christian Solders*.

Geiger counted particles for five years, then left Rutherford's lab in England to teach in Germany. It no surprise that there he perfected an automatic way to count these particles -- the Geiger counter of today. Instead of using a screen that flashed light, he passed the alpha particles through a gas that created a current, which he then amplified into the familiar clicks we heard today.

Today his counters are in demand partly because of fears that terrorists will set off what's called a "dirty bomb." That's not a nuclear bomb, but a conventional explosive loaded with radioactive contaminants, which are dispersed when the bomb goes off. The feeling is that if you have a Geiger Counter you can get enough notice to take an antidote or flee the radiation.

This isn't the first time, though, Geiger Counters have become popular. You can chart the tenseness of our country's mood by the sales of Geiger's Counter. They were bought in large quantities when the Three Mile Island nuclear power plant released radiation in 1979. And during the Cuban Missile Crisis in the 1960s fallout shelters were stocked with them. And of course, they came into prominence in World War II right after the atomic bomb was dropped. Our response, then, was different: Geiger Counters became part of a love song. Here is Doris Day, in 1949, singing the *Geiger Counter Song*, in which she compares her love to the tic, tic, tic of a Geiger Counter.

"I tic, tic, tic, why do I tic, tic?
What amazing trick makes me tic, tic, tic?
I tic, tic, tic, an electric tic
When I feel a realistic tic.
You're such an attractive tic,
You give me a radioactive kick;
It's distractive the way you stick,
But love, love makes me tic.
I tic, tic, tic, and my heart beats quick, How can anything go wrong?
When I'm list'nin' to that Geiger Counter song,
Ya tic, tic all day long."

GPS

I HAVE LOCATED my Diary Queen, it's at exactly 40 degrees 6.986 minutes North and 88 degrees and 13.178 minutes West. I know this because I've bought a GPS, or Global Positioning unit. Although extremely high tech, it works exactly like old fashioned navigation: Locate three fixed points and triangulate your position. Of course for the GPS, the fixed points are twenty-eight satellites orbiting the earth.

A GPS unit measures its distance from these satellites using time. Each satellite sends out a signal containing its position and the time the signal was sent. When it arrives, the hand-held receiver measures how long it took to travel, then uses this to calculate its distance from the satellite. And then it reports its position on the globe with an accuracy of about ten feet.

It amazes me that this degree of accuracy can be purchased for about one hundred dollars, the cost of my unit, although for the GPS system to arise it did take Sputnik, twelve billion dollars, and a war.

In 1957, the Soviet Union launched Sputnik, the world's first satellite. One American engineer, Frank McClure of Johns Hopkins, noticed that he could locate Sputnik by measuring the radio signal the satellite emitted. This gave him the idea for a navigation system based on satellites. Over the next twenty years, the U.S. Military developed McClure's idea into the global positioning system we have today, but it took twenty more years before civilians could use it.

GPS caught the attention of the world during the Gulf War of 1991. The military found it was the only way to move across hundreds of miles of trackless Iraqi desert. This use made huge headlines, and interested the public in the few commercially available GPS devices. But oddly, the devices lacked the main element needed to become really useful: Accuracy.

You see, the military fuzzed out the signal on purpose. They added a billionth of second to the clocks, making the GPS no more accurate than about 300 feet. GPS sales soared, though, when, with little fanfare in May 2000 the government shut off this jamming.

Like all powerful technological innovations, serious uses are being found for GPS almost daily, although the best measure of the acceptance of a technology is its use for pure folly.

The latest fad with GPS is called "geocaching." It's a kind of high-tech easter egg hunt. All over the world, players hid caches—usually a log book and new batteries for the finder's GPS—and then post the longitude and latitude to a web site. Players then hike for hours and endure physical hardship to find the treasures now hidden in at least thirteen countries.

Rest assured that GPS is being put to better uses. So far, a self-guiding plow has been developed, and in the works are talking GPS units that would guide a blind person.

Phillips Screws

A FEW DAYS AGO I struggled with a Phillips Screw. As many homeowners know, they are difficult to remove because the screwdriver tip can spin around in the screw head. You might wonder how such a useless thing got into our homes. The answer is this: They were never supposed to be there.

In 1936 Henry F. Phillips invented, as the title of his patent proclaims, a "Means for uniting a screw with a driver." The essence of Phillips invention is this: Get the screwdriver tip on the screw head, and the driver centers itself neatly on the screw. This means, of course, that the tip also easily pops out of the screw, something many homeowners have experienced. Phillips, though, never intended this for home use—he invented the screw for the auto industry. Phillips knew that car makers needed a screw that could be quickly met by an automatic screw driver, one where the screw driver centered quickly and easily. An ordinary slotted screw and screwdriver were too difficult to line up quickly for mass production.

But in marketing his screen Phillips faced the problem that the nation's biggest screw makers weren't interested. One of the largest manufacturers wrote him: "The manufacture and marketing of these articles don't promise sufficient commercial success to warrant interesting ourselves further."

It took Phillips three years, but he finally persuaded a manufacturer, the American Screw Company, to make the screws. So controversial were they, though, that the engineers there refused to work on them until the company's president threatened to fire everyone. They took a dim view of spending half a million dollars to develop a manufacturing processes for this new, untested screw. Yet, so important was this new screw to the president that he had the first screws off the assembly line plated in gold and silver, and made into a necklace for his wife and a set of cuff links for himself. American Screw then sold the Phillips screws to General Motors, who used them on the 1936 Cadillac. After that they were rapidly adopted: By 1940 nearly every American automaker had switched to Phillips screws. The dominance of the Phillips screw was settled in the 1940s when World War II broke out. Detroit need to produced jeeps and other vehicles quickly, and they turned to the Phillips screw to increase the efficiency of their production line.

In fact, the Phillips screw became so popular that other manufacturers made unlicensed knock offs. The Phillips company didn't press their patent rights, so in 1949 Phillips was stripped of his patent. No one knows exactly what happened to Phillips after this -- he died in 1958 having left the company, and it isn't clear if he died a pauper or a rich man. But his screw has done pretty well. From something that "didn't show much commercial promise" it has become the most manufactured screw in the world -- even going into space: Phillips screws held the space shuttle together.

Saxophone

WE ASSOCIATE THE SAXOPHONE with the mellowness of jazz, but it roots are very hotheaded.

The saxophone evolved from a bass clarinet built by the Belgian Adolphe Sax. It was the result of a scientific approach to designing instruments. For example, the most pressing problem of the time was placing the key holes in a wind instrument so that it was perfectly chromatic. Others did this by trial and error; Sax, though, used mathematics and acoustics to calculate exactly where to place the holes.

He thought so much of his new clarinet, that when he learned a musician in Paris was also debuting a new instrument, Sax dashed there to prove his better. He succeeded, even getting the musician's wife to say to her husband "when Mr. Sax plays, your instrument sounds like a kazoo."

The fiery young inventor returned to Brussels and entered a music contest. They only awarded him the second prize. This enraged Sax, who stormed back to Paris.

He was destitute until the French composer Berlioz learned about Sax's instrument. Berlioz wrote an article praising Sax and overnight the obscure Belgian instrument maker became famous. The fame brought him a loan, which let him start an instrument shop in a dilapidated shed. The temperamental Sax immediately upset the Paris musical establishment. He insisted that every piece for his instruments be made in his shop. At the time, instruments were made piece by piece by craftsman across Paris, then reassembled in a central shop.

So, repellent was Sax to the musical establishment that his shop was often robbed of expensive tools and irreplaceable plans for new instruments—and there were even three assassination attempts on

Sax.

Unable to battle the sabotage any longer, he set up a new shop in a prison, where he trained convicts to build instruments. Under the watchful eyes of the guards there was no longer any smuggling of plans, or sabotage.

During this turmoil, he created his masterpiece: The Saxophone. It was well-received by the public and composers, but its acceptance was blocked by musicians, who threatened to strike if it were used. So, intense was the opposition to the Saxophone that its enemies legally banded into a group called the United Association of Instrument Makers. They sued Sax to strip him of his patents.

The lawsuits bankrupted him by 1852. He disappeared from public view for thirty years until a Paris newspaper reported that he was destitute. And it quoted him as desiring a "few hours [of] peace in a life eaten up by worry." His few friends petitioned the French government, which awarded him a small pension.

Only with his death did the lawsuits that had plagued him for fifty years end. His saxophone, of course, lived on, taking on new life only when jazz entered the musical world.

Technology & 911

W E CAN DRAW MANY LESSONS from the World Trade Center attack of September 11TH, but one thing it clearly reveals is the fragility of the incredible, nearly invisible, web of technology that surrounds us.

The attack severed the web of airplanes, antennas, and cables that connect our world. Packages stopped moving, people stayed put, and telephones were silenced. We've learned by its absence how this technological mesh makes our lives easier, but paradoxically we now see that it makes terrorism easier also.

At first blush the attacks seem very low tech. We expected a rogue nation to lob a crude nuclear bomb, instead we got a relic from World War II: A kamikaze driving a jet. But make no mistake about it, our high tech web enabled every aspect of the attack on the World Trade Centers. Their weapon, of course, was a technologically sophisticated jetliner, but the enabling goes even deeper.

As an example, consider the central role of the computer in training the pilots. It isn't an easy thing to fly a jet into a building. So, the terrorist practiced on the computer flight simulators used today to train commercial pilots. These machines duplicate the cockpit of a jetliner. So exact are these simulations that often a pilot's first flight in a real, live jet is with paying passengers. They allow a pilot to mimic taking off, flying and landing a jumbo jet. Pilots can practice navigating around urban areas because the simulators reproduce the topography of every major city in the world. But the simulators also let two terrorist pilots practice, again and again, crashing into the World Trade Centers. All without the necessity of owning a jet.

The lesson from this is not the evil of technology. The attack reveals the truth of the aphorism "technology is neither good, nor bad; nor is it neutral." It depends on the human at the controls. But the attack should caution us from looking for some quick technological fix to the problems of terrorism.

We want some ultra-sensitive metal detector that will unmask any weapon, or some new x-ray machine that exposes all dangers. Yet, the only lasting solution will be a human one: At the basest level an alert and thinking person at an airport, and at its deepest, agreement and understanding among peoples.

Rolodex

ALTHOUGH WE LIVE IN A disposable society, there are a few things designed so well that they are never out of style. The premier example may well be the Rolodex.

The inventor of the Rolodex, Arnold Neustadter, was fascinated with what he called, in his quirky way, "dexes." By *dex* he meant a method to store and organize information. He used *dex* as a suffix, which he appended to just about every word in our language.

For example, his first invention, in the 1930s, was the *Autodex* a flat phone directory that popped up at the desired letter. Then he moved to devices that helped people create information—for example, the *Swivodex* a spill-proof ink well; and then to the *Clipodex* a writing pad that attached to a secretary's knees. In spite of these failures, he kept thinking of ways to organize information. His big breakthrough came when he modified another of his inventions, the *Wheeldex* to come up with the the classic Rolodex.

Neustadter was trained as a journalist, so he worked with an engineer to design and perfect his ultimate *dex*. They created a masterpiece, partly because Neustadter was very interested in the arts. He collected glass paperweights and studied modern art, and this influence is clear in his classic Rolodex. He wanted it to fit the hand perfectly, be easy to move, and yet be attractive.

The result was simple and elegant: His classic Rolodex has a shiny, tubular steel frame, which is strong and robust, yet streamlined. It's in perfect balance as the card wheel cantilevers over the base. Since Neustadter had a great sense of human kinetics, the knob fits a hand perfectly, and the balanced wheel rotates easily throughout 360 degrees. Neustadter started marketing his new Rolodex in the late 1950s. Its sleek no nonsense design fit well with the efficient image of a 1950s office. Rapidly the Rolodex became a fixture in offices, especially after Jack Lemmon used it in the popular movie *The Apartment.* By the 1970s, it was a true cultural icon, even an emblem of power, growing into an essential tool when the "networking" fad of the 1980s hit.

I've said it was a perennial design, but what about this decade -- surely the computer has forced the Rolodex to the trash heap? No. Some ten million units are still sold every year, and in fact the computer has been modified to accommodate the Rolodex. You can buy programs that will print out cards, which you can store in your old-fashioned revolving Rolodex.

Iridium Satellites

YESTERDAY I LOOKED UP in the sky and saw a flash of light that's best described as "technically sweet."

This odd phrase comes from J. Robert Oppenheimer, a father of the first atomic bomb. He used it to describe projects whose technical challenges were so seductive, and whose solutions were so satisfying that they completely absorbed scientists and engineers—in fact, absorbed them so much that they lost sight of whether the project had any benefits.

That "technically sweet" flash of light I saw was the sun reflecting off the metal antenna of an Iridium Satellite. Its one of sixty-six satellites used to create a now defunct global phone service called the Iridium Project.

The idea was this: Using a special type of mobile phone, a person at any point on the Earth—the ocean, the desert, a rain forest, or Mount Everest—could call any phone on the globe. So revolutionary was this idea that it was hailed as the "first pan-national phone network", and called gushingly "a breakthrough in transcending all geopolitical boundaries."

The Iridium project started when a team of Motorola engineers fell in love with the "technically sweet" challenges of making a satellite phone network. They got five billion dollars from Motorola and other investors to pursue this dream, but within a year or two after launching the satellites, the Iridium Project was bankrupt. What the engineers forgot was this: Who wants a phone that cost three thousand dollars, and can't be used indoors. So, when Iridium marketed their phones to business executives they found few takers. Most, of course, already had cell phones that worked just fine, and few really needed to conduct business from a desert or the top of Mount

Everest.

Although the Iridium company was bankrupt, their satellites are still there for us to see. At just the right time of day, you can catch that flash of light from a satellite: When its surfaces are angled to the sun, it sends a burst of sunlight to the earth. The trick is to know when to look.

To find out, go to the web site *www.heavens-above.com*. If you type in your location, the site will tell you exactly where to stand, what direction to face, what angle to tilt your head toward the sky, and exactly what time to do this—all so you can see a five billion dollar failure. Although I enjoy looking for the satellites, every flash fills me with with trepidation: May my failures never be so public, and may there never be a web site dedicated to instructions on how to view them.

Bathtubs

I'VE STUDIED CATHEDRALS, TEMPLES, and palaces, and even seen in the middle of London a section of an ancient Roman wall. It makes me wonder: What American engineering achievement will be here eons from now. One day while sitting and thinking I came up with the answer: Bathtubs!

In bathtubs, the past is already with us; they aren't far from their ancient moorings. Tap the glassy surface of your bathtub and you've touched an ancient art—the art of Mycenaean jewelers, who specialized in bonding glass and metal together as in a King's specter. An engineer uses the same process to attach a white coating to a bathtub's metal inner shell.

The story begins with the Minoans on Crete. They had bathtubs, but they used them for the Dead, or maybe the Queen, archaeologists aren't sure. The Mycenaeans drove out the Minoans, and took over their bathtubs. They left ninety years later, without the bathtubs. They probably didn't see much sense in washing dead people or queens. Yet soon the bathtub caught on when the Romans began hanging out in huge marble tubs. They felt bathing was a basic social obligation. This pearl of wisdom shows, no doubt, why the Romans founded western culture.

But this idea slipped away until by the 18TH century bodily care sank to total neglect. It had to do with the Counter-Reformation and the Reformation -- they both felt nakedness was a sin. At the end of the 19TH century, after eons of stinking, everyone decided to have a private bathtub. Here is where engineers enter.

They asked: how do you democratize the bathtub? That is, "how do you make a bathtub so that everyone can afford one?" They first made a porcelain tub. But its great weight made it expensive to move, and required a house with extremely strong floors.

In short, the porcelain tub was a Rolls-Royce of bathtubs. What was needed was the Yugo of bathtubs. So, they tried wood, but think of the bottom of a ship—barnacles and the green stuff growing on it. Now think of a week of bathing in a wood tub.

Next they tried metal. This was lightweight, but the tub rusted. So, they painted it, but after a hot bath the paint stuck to people.

Finally, some engineer realized that you needed both a porcelain top and the lightness of metal. The solution: make a thin and light tub of metal, and cover it with a thick layer of frosty glass—the surface you see on your bathtub. And here we return to where we began: Mycenaeans putting glass on metal.

So the next time you desire to see the glorious monuments of history—cathedrals, palaces, temples—instead just take a look at your bathtub.

Tupperware

L ISTEN CAREFULLY, when you open a Tupperware container
you'll here a "burp." That sound is heard millions of times a day.
Here it is. Again.

Tupperware began with Earl S. Tupper, once described as a "failed
tree surgeon and archetypal reader of the *Reader's Digest*." Who, also
"combined faith in technological progress." Yet had a "Protestant
conservative's suspicion of bar-hopping men and cigarette-smoking
women." Tupper was also a prolific inventor of bizarre gadgets: A
"knee-action Gypsy Gun", which explosively removed gypsy-moth
eggs from trees, and self-standing toothpaste and shaving-cream
dispensers.

Tupper spent a couple of years in the 1930s as an engineer at
DuPont, where he came across a new plastic called polyethylene. His
inventive and oddball mind saw many uses for this plastic. So, in 1939
Tupper talked his bosses into selling him a few tons of the stuff
cheaply.

Then, on a farm in a tiny Massachusetts town, Tupper pioneered
thermosetting plastics: By heating the plastic he made it pliable and
moldable. The trick was to find just the right amount of heat so the
plastic became like chewing gun and could be slapped into shape, but
not so much heat that the plastic melted.

After three years of trial and error Tupper finally made a simple
tumbler, but sales of it and other items were poor. To really become
part of our national culture, as it did in the 1950s, Tupperware needed
to tap deeply into our traditions and anxieties. Subliminally
Tupperware resonated with America's fears in the 1950s because it
kept food from rotting in fallout shelters. But tapping into America's
traditions was beyond Tupper himself, so in 1954 he hired Brownie

Wise—a woman with an intuitive grasp of female popular culture. Wise, a single mother from Detroit, pioneered Tupperware's famous "Hostess Parties"; where a hostess, usually a housewife, gathered a dozen or so women in her house and demonstrated Tupperware. In these parties Wise grafted the selling of Tupperware onto the American traditions of sewing circles and quilting bees. And she tapped into the psyche of the new suburban housewife: The parties took advantage of the isolation and neurotic sense of social obligation among suburban women in the 1950s. A feminist called these parties a "form of organizational parasitism analogous to ... colonialism which [used] ... the existing tribal structure ...." But within this "tribal" system Brownie Wise was so successful that by 1954 Tupperware sales topped fifty four million dollars—and made her the first woman on the cover of *Business Week*. And today there is a Tupperware party somewhere in the world every two point two seconds.

Diapers

SHOULD ONE USE DISPOSABLE DIAPERS or cloth diapers? It seems a simple question with an obvious answer: disposable diapers create waste, while cloth diapers are recyclable. But the questions isn't simple at all.

To determine whether a disposable or a cloth diaper is more environmentally friendly calls for an accounting of all of the diapers impact on the environment, from its creation to its disposal.

I can't answer definitely whether you should use disposables or cloth diapers. I can only give you a feeling for what might make a disposable one more environmentally friendly than a cloth diaper and vice versa.

It may seem like I'm defending disposable diapers, but they are the underdog here, and I like an underdog.

First, let's consider the energy used. For disposable diapers, the bulk of energy comes from their manufacture, that's obvious. Cloth diapers, though, also use energy. Growing the cotton, for example, requires farm equipment powered by gasoline and pumps for irrigating with water. And, of course, lots of energy is used in heating the water for washing the diapers.

What about toxic waste? Clearly, disposables contain plastic, and when buried or incinerated can turn into toxic chemicals. A cloth diaper, at first, appears to produce no waste like this, but remember that they're made from cotton and that pesticides are used to grow the cotton. Also, they're washed with a detergent, made, of course, from chemicals.

This may all be true, but surely disposable diapers are clogging our landfills. They aren't really: Disposable diapers make-up one and half percent of the volume of a landfill. There are far worse things filling up our landfills.

So, how do you balance all of these things: Does a disposable diaper use more energy in its lifetime than a cloth one? Does it produce more waste?

I have on my desk studies that claim one or the other is better, but we may never know which diaper is more energy efficient, and less toxic. And I'm not sure that that's the ultimate point.

When I look at where these studies agree, my conclusion is this: We shouldn't be battling about diapers to save the environment. In the scheme of things, they just aren't that important. For example, regardless of which diaper is more energy efficient, the total energy used is small. The amount of energy needed to create a year's worth of diapers for a child is about fifty-three gallons of gas. That's the amount needed for the average weekly commute in an SUV.

So, in debating about what kind of diaper to use we may well be fighting the most convenient battle, rather than the most pressing one.

Thomas Midgley

TWO OF THE GREATEST environmental threats of the twentieth century—leaded gas and freon—were the work of a single man named Thomas Midgley. Because of these two inventions, an historian described Midgley as having "had more impact on the atmosphere than any other single organism in earth history."

In 1921 Thomas Midgley worked as an engineer in the General Motors research labs. His project was to get rid of engine knock— that's when the fuel doesn't burn evenly and makes the motor rattle. Engine knock wasted fuel, something scarce at the time.

Midgley knew the solution was some kind of additive to the fuel. So, he went on what he called a "scientific fox hunt." It was a method he developed as a young man. In school, he and a teammate decided to find out what would make a spitball curve the best. They carefully tested substances, until they came across slippery-elm bark. Midgley used a similar process to find his fuel additive. He systematically tested compounds until he hit upon tetra ethyl lead.

This compound met his goal of increasing fuel conservation: With his lead additive, two gallons would go as far as three gallons without an additive. In fact, despised as lead gas is today, over the twenty-five years we used it, leaded gas saved about a billion barrels of oil.

Of course, we now know the toxic effects of putting lead into the environment.

After this success, Midgley worked on improving refrigerators and air conditioners. At the time, a refrigerator was a dangerous thing because it used toxic, flammable chemicals. Midgley, using the same "scientific fox hunt" as before, found a replacement for these nasty chemicals: CFCs, or chlorofluorocarbons, what we now call "freons."

Today, it strikes us as the most toxic thing in the world. Yet its lack of toxicity and flammability is what attracted Midgley to it. To demonstrate its safety, he once took a mouthful of freon and blew it at a candle. It extinguished the flame showing that freon was neither poisonous nor flammable.

Of course, we learned in the 1970s that freon thins the ozone layer, the essential sun screen that lets life on earth thrive.

There is great irony in Midgley's career: He set out to improve our world, and yet as that historian said he "had more impact on the atmosphere than any other single organism in earth history."

This irony was reflected in his own demise. In 1940 he contracted polio. He designed a system of ropes and pulleys to help him in and out of bed. One day he became entangled in this network of ropes, and was strangled by his own ingenuity.

Glo-sheet

THIS HALLOWEEN I set out some glow-in-the-dark decorations, and they reminded me of a charming story of invention.

In the early 1970s, Becky Schroeder waited in the car as her mother shopped. Becky, a ten-year old, wanted to finish her math home work, but because it was growing dark she couldn't see her paper. She didn't have a flashlight, and didn't want to open the door because the whole car would light up. "I thought," she said later, that "it would be neat to have my paper light up somehow." And with that Becky decided to invent glowing paper.

First, she tried to use fluorescence, but she found that fluorescent things glow only when struck by light—like a black light—so they won't work in the dark. Next she thought of fireflies, but bio-luminescence didn't seem to be something she could incorporate easily into a piece of paper. Then she recalled the glow-in-the-dark Frisbees in her toy box. She learned that they used phosphorescence. That means they store up the energy from a light, then radiate it as a glow in the dark after the light is switched off.

The next day she went to the hardware store with her father and bought some phosphorescent paint. That night she worked in her darkened bathroom, painting stacks of paper, even screaming at one point "It works! It works!" But she found that it wasn't the paper that needed to glow, it was the clip board. She coated her clip board with glowing paint, so that it gave off just enough light to shine through the paper on top of it. Over the next couple of years—from age ten to twelve—Becky refined her device to use batteries to generate the glow. Then she patented her "glo-sheet", becoming the youngest female inventor in U.S. history.

Written up in the *New York Times*, the devices sold briskly. Photographers used them in their darkrooms, critics found them perfect for taking notes in dark theaters, and emergency medical technicians used them in ambulances.

NASA even came knocking at her door. They'd been developing something similar, and noted that if she were a former employee the patent would belong to them. Of course, at age twelve she wasn't a former NASA employee.

You might use Becky's story as inspiration for your own inventions, but I'd recommend staying away from glow-in-the-dark stuff. Since Becky, people have invented golf balls that glow, although why you'd golf in the dark is beyond me; one man sells a whole bicycle covered in glowing stuff, and there is glow-in-the-dark toilet paper, invented so you can find it at night. It's been discontinued, though, because after you use it your bottom glows.

Numbers & Empire

A S YOU CHANGE YOUR CLOCK for daylight savings time, take a good look at it: Its the tool Western Europeans used to conquer the world.

A clock? Conquer the world? Yes. The power to measure time helped turn the backward 9TH century tribes of Europe into the powerhouses of the world. Europe trailed other regions badly in the 9TH century: Western Europeans ruled the world—Portugal, Spain and the Netherlands had huge empires. And by the 19TH century Western Europe's domination reached its apex in Queen Victoria's empire.

How did these backward 9TH century Europeans accomplish all this? The obvious answer is "science and technology", but I've learned from historians that there is a more specific answer—one that warms my engineer's heart—and that is numbers.

The West brought together mathematics and measurement to record reality, and thus the power to control it. And this brings us to clocks. Clocks were the first way Europeans quantified the world. The chime of the town clock chopped the day into numbered segments; calling out the time to start or stop trading, or go to church. This was a sharp contrast to days marked only by dawn and sunset. Quantification spread to all aspects of life.

Numbers affected music, armies, art and navigation. The free form Gregorian chants of the 9TH century gave way to music with a rich meter controlled by a clock. And it was a short step from regimented music to powerful Armies. The political philosopher Machiavelli noted this: Just as a dancing man keeping "time with the music, cannot make a false step; so an army that properly observes the beat of the drums cannot easily be disordered."

And mathematics allowed people to divide space into numbers, giving them maps overflowing with compass bearings, tide tables, and even the times pirates might be expected. These number-laden maps guided sailors across the seas to conquer new worlds. Bookkeepers armed with numbers followed the sailors. They used double-entry bookkeeping to control commerce, industry and government. Double-entry bookkeeping doesn't sound like a world-changing event, yet it allowed a merchant to "picture" the reality of his or her business.

The poet Auden summed up the result of all these numbers for the West: We live in societies "to which the study of that which can be weighed and measured is a consuming love."

Not to me: Tomorrow my alarm clock will screech and command me to divide my day into bits and pieces, but when I rise, I probably won't feel like following my Western heritage and conquering the world.

Nitrogen Fixation

THE MOST IMPORTANT technical achievement of the 20ᵀᴴ century may well be that of Fritz Haber and Carl Bosch. They found a replacement for, well, the polite word is "bird droppings."

They replaced *quano*—that's the name for the nitrogen-rich excrement of seabirds—which 19ᵀʜ century farmers used to fertilize fields. "Nitrogen-rich" is the key.

We need to eat nitrogen-containing amino acids for our bodies to grow. If the soil is rich with nitrogen, plants will absorb it and make these amino acids, which we can then digest.

The problem with using guano was that it was non-renewable. For nature to generate large piles of quano requires a place with huge numbers of nesting birds, abundant fish stocks for these birds, and a rainless climate.

Islands near Peru are one of the few places that meet these conditions. Nineteenth century merchants leveled the guano cliffs there, transporting tons to the farmers of Europe. By 1870, with the guano piles gone, a desperate search began for a replacement.

They turned to the air. Air is filled with nitrogen, millions of times more than any human being would need in a lifetime. But, you can't just put air in the ground. You must remove the nitrogen, combine it with other elements to make a liquid, which can then be added to a field.

For years chemists tried to remove nitrogen from the air, but it wasn't until a German chemist, Fritz Haber, found the secret: extremely high pressure. He found that he could use the nitrogen from the air and form ammonia if he did his chemical reactions in a vessel pressurized to 100 times normal atmospheric pressure. But using these high pressures presented great problems with making large quantities.

That kind of pressure can be contained on a small scale, but make a vessel large enough to make tons of ammonia and pressurize it to 100 times atmospheric pressure, and you have a huge bomb if the vessel fails.

This is where Carl Bosch enters the picture. When he learned of Haber's process in a meeting, Bosch said "I believe it can go. I know exactly what the steel industry can do." Using this knowledge he designed vessels to contain the pressure and created a factory that made ammonia by the ton.

So important was this work that Bosch won the Nobel Prize for Chemistry. The citation called his ammonia factory "a technical advance of extraordinary importance." Indeed, nitrogen from the Haber-Bosch process provides the nutrition for about 60% of the world's people.

Although making ammonia to supply nitrogen doesn't sound as exiting as space travel or the latest computer, I'd still nominate the work of Haber and Bosch as the most important technology of the last century. Feeding the world, after all, is a pretty big thing.

Jell-O

O N YOUR HOLIDAY TABLE there is product of one of the greatest American engineers of all time. In 1845, Peter Cooper invented Jell-O.

Cooper had a hand in most of the amazing achievements of the late 19TH century. He designed the first American locomotive, and helped lay telegraph cables under the Atlantic ocean. He also owned a glue factory, which led to Jell-O.

Glue, at the time, was made by rendering animals. Rendering means, literately, to melt down an animal. There are always rumors that Jell-O is hooves or other disgusting parts of an animal. In a way it is.

Jell-O is made from a protein called collagen. In mammals, like us, it makes up about a third of our bodies, and makes our skin, bones, and tendons strong and elastic. The collagen in Jell-O comes from cows and pigs, but it goes through so many steps of processing that the FDA doesn't consider it a meat product; in fact, Jell-O is the only pork product certified Kosher.

Cooper figured out how to render animal collagen so that it turned into the flavorless, wiggly clear substance we now call Jell-O. He advertised it as "a transparent substance containing all the ingredients fitting it for table use ...." But few tables took up Cooper's invention.

It resurfaced a half century later when a man with the odd name Pearl B. Wait took it up. Pearl Wait and his wife made two great innovations in Jell-O. First, they added flavors—strawberry, raspberry, orange and lemon.

And second, his wife added the "O" to Jell-O, basing it on a product their neighbor sold, a coffee-substitute called Grain-O. The "O" ending was a fad at the time, much like "-o-rama" was in the

1950s.

Wait failed, though, like Peter Cooper in getting any interest in his Jell-O. So, he sold it to their neighbor who made "grain-O."

This neighbor, a Mr. Woodward, placed ads featuring actress and opera singers, suggesting that Jell-O was a dessert of the elite. He hired Norman Rockwell to illustrate the ads. And he sent fleets of stylishly dressed salesman out in handsome, horse-drawn carriages. They demonstrated Jell-O at fairs, picnics, teas, weddings, and church socials. By 1902 Woodward was a millionaire.

Jell-O went through a golden age in the 1950s, then sales declined in the 60s and 70s. Lately, though, it has come back with a vengeance. Jigglers—a high-density Jell-O finger food—is popular with preschoolers, and a vodka-laced version appeals to the young adult party set.

I think, though, there is a fundamental reason that Jell-O will always be with us. It's this: Jell-O is like us. I've learned that if you hook up an EEG Machine to lime-flavored Jell-O, the Jell-O shows a pattern matching a healthy adult's brain waves.

Frisbee

THE FRISBEE BEGAN as something called the "Pluto Platter", and it would have stayed that way except for two things: The fanaticism of Ed Headrick, and the demise of the Hula Hoop. In 1964, Headrick asked the Wham-O toy company for a job, but they told him they were short of money. Headrick, though, was so interested in developing toys that he agreed to work for free for the first three months.

He became fascinated with Wham-O's Frisbee, then called the Pluto Platter. As he once said "I felt the Frisbee had some kind of spirit involved. It's not just like playing catch with a ball. It's the beautiful flight."

Just as Headrick joined Wham-O, the hula hoop craze ended. This left the company with a warehouse of unsold hoops. Headrick realized that if he melted down the Hula Hoops, Wham-O could make more of these flying discs. But before he could do this he needed to improve them. The disks sold poorly, partly because they wobbled too much as they flew, dropping out of the air after only a short distance. Headrick made the crucial improvement that made the Frisbee really fly. He cut small grooves in the top of the Frisbee. He described their purpose in his 1967 patent: These grooves "exert an interfering effect on the air flow over the implement [the Frisbee] and create a turbulent unseparated boundary layer over the top of the implement reducing aerodynamic drag." In other words, it makes the air stick to the top of the Frisbee, letting it fly farther.

Next he got rid of the name Pluto Platter. Wham-O had chosen the name to appeal to obsession with outer space and flying saucers in the 1950s.

Under Headrick's guidance, all references to planets and UFOs were dropped. He thought they seemed childish, and he was right. Students, in particular, flocked to buy the faster, cooler frisbee. Sales of these so-called Professional Frisbees zoomed to over 100 million within thirty years.

Headrick never fell out of love with the Frisbee. In 1976, he created the sport Disc Golf where players hurl Frisbees at metal cages, which is now played by over two million Americans. He dedicated the remainder of his life to promoting this game, claiming that "Disc Golf is my life." He saw the game as a "new way", in his words, for "wandering people who have graduated from High School with no purpose in life." He traveled around America to give demonstrations of the sport he loved.

He joked once, "We used to say that Frisbee is really a religion -- 'Frisbyterians,' we'd call ourselves." Adding, "when we die, we don't go to purgatory. We just land up on the roof and lay there."

And when he died, Ed Headrick requested that his family have him cremated, and have his ashes molded into Frisbees.

MUZAK

M Y WIFE, AMY, AND I SAT outside our favorite restaurants waiting to be called for brunch. As we entered, we heard something that G.K. Chesterton, the great British journalist and grump, called the worst feature of modern life. He called it "monstrous and ominous", and said it indicated "moral chaos."

We heard Chesterton's monstrous noise: A MUZAK version of Simon & Garfunkel's "Feeling Groovy." Chesterton's "moral chaos" was music that was piped in while eating at a restaurant. Yet, to me this is MUZAK to my ears, because it's the work of a fellow engineer: George Squier.

To invent MUZAK, Squier fused, in the 1920s, long distance telephone service and radio technology.

He invented a way to broadcast radio signals down a telephone line, but let the telephone calls still take place—something called carrier transmission.

To capitalize on his invention Squier created Wired Radio, a nationwide broadcast of music and public service announcements.

Squier used one central station to create his broadcasts, then transmitted them via phone lines across the nation to his stations, which broadcast them locally. He charged customers eleven cents a day for the service, yet Squier's company nearly failed.

When he patented his invention, he chose to allow any U.S. citizen to use it. This let AT&T, which dominated the phone lines, use the technology to help others compete with Squier's Wired Radio.

To overcome his rivals Squier invented a brand name for his broadcasts. He combined the word "music" with the most popular trade name of the time, Kodak, to come up with MUZAK. Squier thought of MUZAK as a mix of news and dance music, but when he retired, his successor, a Wall Street banker, changed the format.

The banker learned that Westinghouse and General Electric used music to increase their workers output, so he repackaged MUZAK to be piped into factories—and eventually into every nook and cranny in America. MUZAK was even played in the Apollo Thirteen space module during its time of troubles.

Today MUZAK is still around with eighty million listeners. And what of Major George Squier's method of putting two signals down a single wire—that carrier transmission?

Well, in addition to the music I hear right now, its also used today to bring 500 television channels into your house on a single cable.

Garbage

THIS MORNING I'm taking out the hallmark of human presence: Trash. We've found two thousand year old waste from Mayans; and the first time humankind left the earth—to visit the moon—we left our urine bags behind. But I'll stay closer to home today, I'm going to take a look through my trash as I wait for my trash truck.

I'm looking for fast food packaging, diapers, and Styrofoam because most people think dumps are overflowing with these things, estimating that they take up over fifty percent of a landfill's volume.

I don't see much of any of this. Yet, what could be more emblematic of our wasteful culture than fast food? The reality is this: Fast food packaging makes up less than one-half of one percent of a landfill's volume.

Now, I see foam products: Egg cartons, meat trays, coffee cups, and packing peanuts—yet all of this adds up to about one percent of the volume of a landfill.

What about diapers, then? The *New York Times* called them the premier "symbol of our nation's garbage crisis." Nearly ninety percent of American babies are diapered with disposables, yet diapers take up a bit less than two percent of a landfill.

So, if the main things in trash aren't diapers, foam containers, and fast food packaging—which most people think fill landfills, yet only take up about three percent—what exactly is this truck going to pick up? As I take a look through here I see some plastic trays and bottles. That's part of the answer. They take up about fifteen percent of a landfill's volume, but this percentage has been decreasing because plastic things are now made thinner than they used to be. So where is the problem with landfills?

What clogs our landfills is paper.

A year's worth of the *New York Times* weighs fifty pounds and takes a volume equal to about fifty thousand Big Mac containers. Landfills are stuffed with paper packaging, paper plates, junk mail, and computer paper. The computer—the foundation of our paperless society—is the root of our paper problem.

It would seem the answer is recycling. About thirty percent of paper is recycled, but that percentage is staying steady. Partly because the recyclers have picked the low hanging fruit—things that are easy to recycle like newspapers and corrugated cardboard. They've got to find a way to cheaply recycle mixed papers, and contaminated papers. So, in the meantime there is one thing you can do to help the landfill problem—it isn't less fast food, or using cloth diapers, it's using less paper.

Waiting in Line

I'M SURE THAT THIS week you've been trapped in a slow moving line, likely for last minute gift shopping, or to return something. As an engineer, I have some good news for you about making those lines move faster. It comes from a branch of engineering called "queuing theory."

It began in 1909 at the Copenhagen Telephone Company when an engineer there, Agner Krarup Erlang, figured out how to direct telephone calls to unused phone lines. This is exactly the same as a check out line. You showing up to check out is like a phone call arriving, and an available cashier is like an open telephone line. Here's what Erlang learned that is relevant to your life.

To keep the lines moving it would seem that the store should just measure the number of people arriving in a typical hour and then assign enough cashiers so that usually everyone will be served quickly. Erlang showed that this was a recipe for checkout line gridlock.

He learned that people are as likely to arrive at one time as another, but precisely when they arrive is random. This means that people will arrive in bunches, not spaced out evenly. So, if stores have just the right number of cashiers for the average number of shoppers in an hour, the store will, at times, have too few cashiers, resulting in long waits. The solution is to combine the separate lines into one huge line, and let that line feed to more than one cashier. You see this used at airports and in banks. It works because there is a random chance you'll be behind someone who causes a delay: Someone, say, who needs a price check. If that delay happens when you're in a line for a single cashier you'll be delayed, but imagine if you were in that one line that fed, say, three cashiers. The only time you'd be delayed is when people in front of all three cashiers are delayed. This makes the single line about three times faster than having one line per cashier.

Erlang's work also answers the pressing question for this season: Why do other lines always move faster than yours? The answer: Because it's true. It's that randomness of delays again.

Picture yourself in a line, with a line on each side, so three lines total. If it's random that someone in one line will have trouble, all three lines are equally likely to suffer a delay, which means that there is only a one in three chance that your line will suffer the least. To put it another way: The chances are greater, two in three, that one of the other lines will move faster.

So, here is what mathematics is telling us: Sure, today that other line is moving faster, but some days you'll be in the faster lane. In other words, during this season: Let the odds be with you.

Henry Dreyfuss

I HAVE, IN MY OFFICE, a black telephone that's over fifty years old. I'm sure you can picture it, it's *the* classic phone. I like it because the handset fits my hand perfectly.

It isn't a surprise that I like its shape. It was designed by Henry Drefyuss, a man who kept very careful tabs on the size of human beings.

Dreyfuss was an industrial designer, who created nearly anything used by a human. As he liked to say, anything that was "going to be ridden, sat upon, looked at, talked into, activated, operated, or in some way used by people." This included, eventually, John Deere tractors, a jetliner cabin, and even the inside of *Time* magazine. He claimed that his ability to design things for humans came from his work in the theater.

At age seventeen he designed sets for a Broadway theater. Theater design, he said, requires a person to visualize and create a mood, yet be practical about placing entrances and exits, and to be considerate of the actors who work in the settings.

After designing 260 sets, he quit the theater and opened an office in New York. With only a card table and folding chairs for furniture, he proclaimed himself an industrial designer. At first, he got only small design jobs—shaving-brush handles, perfume bottles, belt buckles, and neckties. But he used these jobs to refine his approach to designing things for people.

His secret, he claimed, were Joe and Josephine, who had places of honor on the walls of his office. They were austere line drawings of a man and woman, covered with the average measurements of every aspect of a human being: The size of an arm, a finger, or a leg.

He used this information about Joe and Josephine to figure out what a typical person could easily do. For example, he learned the average distance a person can reach without strain, that handles under half an inch in diameter are likely to cut into the hand under heavy loading, and that handles more than one and one quarter inches in diameter feel fat and give a feeling of insecurity. So, when Bell Lab approached him about designing the phone, he used the measurements of 2,000 faces to determine the average spacing between mouth and ear.

There is a place in your home where you're very likely to see his work. On your wall is probably a round, Honeywell thermostat. If so, reach up and appreciate how well its dial fits your hand. That'll be your tribute to Henry Dreyfuss, who, more than any other person, changed the way we feel our world.

Swatch

WHEN THE CLOCK STRIKES midnight and a New Year begins, it's likely you noted the New Year's arrival on a Swiss-made *Swatch* watch. To my engineer's eye, the Swatch is an incredible watch, which may sound odd since the Swatch is famous for being cheap, and mostly plastic. Here's the Swatch story.

In the 1980s the Swiss watch industry was near collapse because the Japanese digital watch had taken over. Deeply in debt, the Swiss banks hired a man named Nicolas Hayek to wind down the industry.

Hayek owned an engineering consulting firm and had advised many companies, like BMW and Porsche, on how to make their products better. Hayek gave the banks advice that they didn't expect: He told them, don't shut down the Swiss Watch Industry, instead retool it to build the world's thinnest, cheapest watch. The banks opposed this idea, so Hayek backed the plan with his own money.

He hired Dr. Ernst Thomke to design the new watch. Thomke had worked as an apprentice mechanic in the watch industry, but had left many years ago to study medicine. He hesitated to leave his secure job and return to a dying industry, but when he saw how much the Swiss watch industry had decayed, he rose to the challenge.

To make a thin, cheap watch, Thomke and his engineers had to reinvent the watch completely. Traditionally, watchmakers start with a watch case and fill it piece by piece with parts, repeatedly flipping over the watch to insert new items. Thomke invented a way to make a watch in one continuous step, no time-consuming flipping needed. He designed a series of robots that created the plastic watch case in one swift move, embedding, in this case, many of the watches parts. This reduced the number of moving parts from from ninety to almost fifty, and lowered dramatically the cost of making the watch.

Thomke's production line makes a Swatch every three seconds—no wonder it's often referred to as "printing" a watch. In fact, so innovative is the process, but so simple the watch, that the patents the Swatch company holds today are not on the watches, but on the robotics.

A Swatch is often cheeky in design, poking fun at the stuffiness of traditional Swiss manufacturing. Yet, it made the Swiss, again, the world's largest watchmakers. Over 200 million Swatch have been sold since Nicolas Hayek suggested the world's thinnest, cheapest watch. Swatches are so popular now that collectors pay a premium for never opened boxes, even going so far as to dust for finger prints.

So proud is Nicolas Hayek of this achievement, he wears at least eight Swatches at a time—covering him from wrist to elbow on each arm. And he can afford these watches. Owning Swatch has put him at No. 156 on Forbes Magazine's list of the world's billionaires.

Fluid Flow

TODAY I'M TALKING to you from a most unusual place. I'm about to hop into my shower and share how an engineer solved one of the great mysteries of our time.

As I shower, the curtain billows into the tub. After years of doubt about exactly why this happens, we now have an answer from one Mr. David Schmidt. Schmidt, an engineering professor in Massachusetts, is an expert in something called computational fluid dynamics.

The motion of many fluids, like water and air, is described by a single equation, called the Navier-Stokes equation. Discovered in the 19TH century, it's so important to our world that there is a million dollar prize for anyone who can solve it completely.

Engineers, though, with a computer can approximate its solution. They use this equation to calculate the flow of air over a jet, or how blood flows through our bodies. Schmidt, himself, is an expert in using the computer to solve problems with spraying fluids, exactly the case in a shower.

There are two competing theories about why my shower curtain billows into the tub. The first is the Chimney effect, which says that the hot water heats up the air in the shower, causing the curtain to rise. The shower stall then sucks in cold air, pushing the shower curtain into the tub. The second theory is based on the Bernoulli effect: that air rushes past the curtain, lowering the pressure along the curtain, sucking it in. This is, by the way, why an airplane flies.

Now, Schmidt programmed his home computer to simulate the flow of water past the curtain. Using nearly $30,000 worth of software, he created a virtual shower that flowed for thirty seconds at about eight gallons a minute. It took almost two weeks of computing, some 1.5 trillion calculations. Schmidt's computer showed him that we create a hurricane in our showers every morning.

The shower pushes the air into a large swirling vortex with a low pressure center. The shower's water droplets decelerate because of aerodynamic drag, transferring their energy to the bathtub's air, which then twists like a hurricane in the bottom of the tub, pulling in the curtain. You can even see this hurricane, Schmidt says, if you blow smoke into the tub.

Now, this might all seem kind of silly, but the study of how fluid drops move is of vital importance. For example, many researchers use Schmidt's computer programs to study the scattering of fluids in asthma inhalers. They want to find better ways to deliver drugs to the lungs. Now, that would be an important problem to solve, unlike the shower curtain. After all, if you don't want to be bothered by a curtain, just buy a door for your shower stall.

Television Remote Control

We have officially been in the couch potato age for almost fifty years. It began in 1956 with Dr. Robert Adler at Zenith. He invented the remote control, called, in that age of Sputnik and rocketry, the "Zenith Space Command." It wasn't the earliest television remote control, but it was the first successful one.

It evolved from a remote called the "flashmatic." This controller flashed light at special photocells installed in the television set. These light signals could turn the television on and off and also rotate the channel selector knob. But it had to be used in a darkened room, because sunlight made the tuner rotate continuously.

Zenith engineers liked the idea of a remote control and so tried to improve on this design by using radio waves. But radio signals travel through walls and so could control a TV set in a nearby room. To find the right invisible signal, they turned to Robert Adler.

Adler was an expert in combining sound with electronics—a field called, not very cleverly, acousto-optical electronics.

Adler didn't want to use sound that could be heard by humans, he thought it would annoy users to hear a beep every time they changed a channel, plus some household noise, or a sound from the television itself, might trigger the channel to change. So he used "ultrasonics." That is, high-frequency sounds that are beyond human hearing.

Adler built a kind of silent "chime" that called out signals to the TV set. He placed inside his remote control four lightweight aluminum rods. These were of just the right size so that when lightly tapped, each emitted a high-frequency sound unheard by humans. He built into the TV set a special electronic circuit that could hear these sounds and react.

Adler's remote control, called the Space Command, but nicknamed the "clicker", was first marketed in 1956—the year when we officially entered the couch potato age. It took a while for this couch "potatoness" to take hold because Adler's remote added nearly 30%, to the price of a TV. But, of course, human nature won out and by 1985 more televisions were sold with a remote than without.

What did the inventor of the television remote control think of how his little device evolved? In 1999 he said, "the thing has so many buttons, I don't know what most of them are for. And," he added, "frankly, I couldn't care less."

Colton

RECENTLY, I PICKED UP a book about spices, although it was really a geopolitical history. It focused on a tiny Indonesian island that became, in the 17TH century, a battle ground for the English and the Dutch, with, of course, the natives caught in the middle. They all wanted the nutmeg that grew there. In the 17TH century, a man could sell a small sack of nutmeg for enough to build a large house and then retire there in comfort. The value of nutmeg came partly because of its rarity, but also because Europeans thought it had powerful medicinal qualities. So, for two centuries the English and Dutch battled over the Island, decimating it in the process.

At first, when I read of these "nutmeg" wars, I thought how quaint that the European economy should depend on spices from obscure parts of the world. Yet, by the time I finished the book, I realized that my own world operates in exactly the same way.

The electronic network that I live in—my computer, cell phone, and pager—depend on something call Coltan. It's as magic to us as nutmeg was to a 17TH century European.

Coltan looks like black mud. It's name is a contraction of columbium and tantalum. And it's that tantalum that's important to our world. A gray-blue, very hard metal, it's the key element used in a device called a pinhead capacitor. These electrical devices regulate the voltage and store energy in cell phones, pagers, and computers. In the last few years alone, tens of millions of these tantalum-filled capacitors were manufactured.

Coltan is found in three billion year old soil, like that of the Rift Valley in Africa, which contains eighty percent of the world's supply. And, of the eighty percent, the majority is in the region.

And much like the nutmeg of the 17TH century, Coltan has brought ruin to the Congo. It has made the area attractive to neighboring countries, and Coltan has been a key force in accelerating the civil war within the Congo. By some estimates, these resource-based wars have killed about five million people, and displaced another ten million or so.

No doubt, some generation after us will evolve past cell phones and pagers, and will no longer need to run their world with the tantalum that comes in the magic mud coltan. And, no doubt, they will look back at the coltan wars and think them as quaint as the "nutmeg" wars of the 17TH century. The message is clear: As we use our cell phones today, we should remember those nutmeg wars, and keep a careful eye on how our technological systems affects the world.

To use George Santayana's aphorism: "Those who cannot remember the past are condemned to repeat it."

Zildjian Cymbals

H ERE IS A SOUND with roots in sixteenth century
Constantinople. That's the sound of a Zildjian cymbal.

It began in 1618 when Mustafa the First, Sultan of the Ottoman
Empire, ordered an Armenian metalsmith named Avedis to create
cymbals. The Sultan wanted them for his for elite guard, which used
them to spur themselves in battle, and to strike terror in their enemies.

Avedis made his cymbals from bronze, an alloy of copper and tin.
It's made by melting the two metals together, although if not mixed
correctly the metals will separate within the cymbal, creating patches
of copper and patches of tin. If this happens the cymbal won't ring
because the sound can't reverberate.

The key to Avedis' success though wasn't the bronze—for centuries
cymbals and bells had been made from copper and tin—the secret was
in his special way of mixing the two metals. It yielded bronze that
held its strength when hammered to unimaginable thinness.

The resulting cymbals so pleased the Sultan that he gave Avedis
eighty gold pieces, and changed his surname to Zildjian "Zil" is
Turkish for cymbal, "dj" for maker, and "ian" means son of. So,
Zildjian means "Son of cymbal maker."

With his new status, Avedis refined his cymbals, expanding into
Greek and Armenian Churches where cymbals were used to
accentuate the hymns and chants.

Many other cymbal makers coveted Avedis' secret methods. To
keep it out of their hands, he passed the special mixing process orally
to his eldest son. By keeping this secret for centuries, the Zildjian
family kept at the forefront of cymbal making, responding to new
markets.

By mid-19TH century opera, with its many themes rooted in ancient myths, adopted the Turkish cymbal.

After three centuries of manufacture in Turkey, the secret Zildjian cymbal formula passed on, in 1929, to the the oldest living Zidjian male heir: An American immigrant also named Avedis Zildjian.

Just like his predecessor Avedis responded to market need. This Zildjian introduced the cymbal to Jazz.

Partnering with Gene Krupa, the great drummer from the 1920s, he produced a cymbal called the "Paper Thin." Its brightness and quick decay livened up jazz, and became the instrument drummers used to keep rhythm.

Zildjians are now the cymbal for popular music. Ringo Starr is said to have used the Zildjian line on all the classic Beatles recordings. And they're favored by Lars Ulrich of Metallica, Ginger Baker of Cream and Mitch Mitchell, Jimi Hendrix's drummer.

Today the Zildjian cymbal still sets the standard, but the latest Zildjian patriarch has deviated from the tradition of passing trade secrets to the oldest son.

Armand Zildjian, who died recently, shared the formula for mixing the metal with his daughter. She's the first woman to know this since the company's beginnings in 17TH century Constantinople.

Silicon

L AST NIGHT, AS I PUT together a PC, I examined the computer's memory chip—in technical parlance a 32 MB DRAM chip.

It seems innocuous: It has no moving parts and yet, because its part of the electronic revolution, will change the world. So revolutionary is the chip, that we've listened to a decade of jabbering about the "new economy."

The implication is that the microchip leaves behind the industrial world of manufacturing—the days of grimy work putting together things like washing machines or cars. No doubt, the microchip is revolutionary, but make no mistake, it is very much part of the old industry, as much as the new—in fact by some measure even more so. Its just hidden from us.

A key measure of the environmental impact of manufacturing something is the weight of fossil fuel required to produce it, relative to the product's weight.

For example, in building a car you need steel, plastics, paint, and so on. So, if you add up all the fuel used to make these parts, and then add in the energy needed to make the car itself, you'll find that for every pound of car produced, it takes two pounds of fuel. Now, compare this to a microchip—in fact, that 32 MB memory chip I mentioned. This tiny, non-moving thing takes an astounding 630 times its own weight in fossil fuel to make. How could this be possible?

Purity is the answer.

A chip is made of silicon, the main ingredient of sand. To work its magic the silicon used must be extraordinarily pure. If you were to shrink to the size of an atom and wander through the silicon used to make this chip, you'd pass by a million silicon atoms before you'd find a single impurity. Any more than that and the chip wouldn't work. Now, to purify sand to that point takes a huge amount of energy—about a quarter of the total energy needed to produce the chip.

In addition, all this work must be done in a clean room. That is, a room which is many times tidier than a hospital operating room. The climate and ventilation of the room must be rigorously controlled. This takes about 50% of the energy used to make the chip.

So, the upshot is this: During its lifetime a microchip uses very little energy, but it takes a great deal to produce it.

For example, if you buy eight new computers over a period of ten years, the total energy used would equal that needed to produce a car.

So, this notion that the microchip makes for a brand new economy is only half right. It's very useful to keep our eye on that other half that is still the old, industrial manufacturing. It reminds us that there is no free lunch.

Why a Chair?

IN THEIR MOVIE *The Coconuts*, one of the Marx Brother's asks "Why a duck?" In the same spirit, I ask today "why a chair." Oddly, I think the Marx Brother's question might be easier to answer, but I'll give it a try.

It seems, of course, that a chair is a natural response to how we bend at our ankles, knees, and hips, but it isn't at all. One thing is for certain: Our use of the chair was nurtured by our response to social, not genetic or anatomical forces.

There is much evidence that people raised in a society that squats are perfectly comfortable and healthy. In fact, it's a natural thing for children to do it, until the chair conditions it right out of them. And there is much evidence that often it's better not to sit. For example, recumbent bikes where the rider leans back are thought safer, speedier, and more efficient.

So, why a chair?

The chair has been around since ancient times, although never an essential part of a household. In Rome, for example, the bed was the all-purpose piece of furniture. Besides sleeping in it, a Roman would eat, read, and write on their beds. The chair made few steps forward for many centuries, even taking some steps backwards. In the 7TH and 8TH centuries, Arab conquerors, a desert people with no steady wood supply, replaced the chair-level ways of the pre-Islamic Middle East with floor-level seating.

It wasn't until the 19TH century that the chair began to dominate among furniture. The Industrial Revolution propelled the chair into our homes. Partly because mass production made chairs cheaper, but also because work itself changed.

Industrial work was more likely to be seated than agricultural work. Work took place at assembly lines and in offices—work often done seated in chairs.

And it looks as if the chair is now unstoppable. Chairs are locked into our architecture: Windows are set in buildings such that we have to sit 18 inches off the ground. And the chair is firmly ingrained in our culture. University professors hold chairs, and we have chairmen, chairwomen, and chairpeople, and as an outgrowth county seats, district seats, and seats on the stock exchange.

In fact, the chair has become so potent a symbol of Westernization that Ghandi, and more recently the Ayatollah Khomeini, were always photographed sitting on the floor rather than in a chair.

Even in Japan the chair is taking over. Many households have both rooms with the traditional mats, and Western rooms with chairs. But now younger people are finding it difficult to sit on the ground, so the mat rooms are disappearing.

By now we are shaping the chair less and less, and the chair is shaping us more and more. Certainly we sit too much since some thirty percent of Americans are now obese.

Hydrogen-Powered Cars

IN HIS STATE OF THE UNION address, President Bush proposed spending over a billion dollars to build a hydrogen powered car. He explained it like this:

> *"A single chemical reaction between hydrogen and oxygen generates energy, which can be used to power a car -- producing only water, not exhaust fumes. With a new national commitment, our scientists and engineers will overcome obstacles to taking these cars from laboratory to showroom, so that the first car driven by a child born today could be powered by hydrogen, and pollution-free."*

Indeed, he's correct to say that a hydrogen-powered car produces no pollution, but he needs to add the phrase "when operated." The key obstacle to these cars, and their Achilles heel, is getting the hydrogen to the car.

It would seem easy because hydrogen is abundant—two-thirds of the ocean is made of the element—yet hydrogen isn't handy. It's always bound to something else, and this means it must be manufactured, and there's where pollution enters. Ninety-six percent of the hydrogen produced comes from natural gas, oil and coal—exactly the fossil fuels we'd like to abandon!

Currently the source of hydrogen is natural gas. Each molecule of natural gas contains four hydrogen atoms bonded to a carbon atom. To release the hydrogen you blast apart the natural gas. To do this requires energy, which, of course, comes from burning either coal or a petroleum product.

This suggests we should look at alternative sources of energy for making hydrogen. Perhaps we could just harvest the sun's power by using solar energy to produce the hydrogen. Alas, this isn't practical yet. To produce an adequate amount of hydrogen we would need enough solar collectors to fill all of Connecticut with photovoltaic cells. Some have suggested instead, putting the collectors in space. That would require putting forty dishes, each the size of Manhattan into orbit. Of course, this would cost a tremendous amount of money.

An alternative is to use plants. They contain hydrogen bound as carbohydrates. Currently this is experimental, but even if successful plants are not a cure-all for our fossil fuel woes. Plants use fertilizer that is made mainly from fossil fuels. Analysis shows only a modest gain in pollution reduction.

So, where will the energy come from? The key is to get the power to make the hydrogen from somewhere other than fossil fuels, and that leaves us with an alternative that many don't find palatable or environmentally friendly. If we don't want to just repackage fossil fuels, we must get it from atomic energy. *That* will be the heart of the hydrogen economy.

Clarence Birdseye

L AST NIGHT I ATE a frozen dinner. I owe that dinner to a man who relished eating seal, whale, and caribou, and who called the front part of a skunk a rare delicacy. His name is Clarence Birdseye. He worked as a naturalist for the government, although I didn't realize that naturalists ate as much as they watched.

In 1912, Birdseye traveled to Labrador, in Canada, to trade fur. One winter day, while ice fishing, Birdseye piled his catch beside his fishing hole. The combination of ice, wind and temperature instantly froze the fish. When he took them home and cooked them, he was surprised they had the taste and texture of fresh food. Four years later he returned to New York, determined to find the secret to frozen food. He could afford to spend only seven dollars for equipment: He got an electric fan, salt, and ice—and borrowed the corner of an ice house from a friend. He discovered that the cells of the fish had been frozen so quickly that there was no time for large crystals to form. It was these ice large crystals that broke the fish's delicate cell walls and killed it. Then on defrosting the broken cells let vital fluids leak out, destroying the taste of the tissue. So, the key to making palatable frozen food was to freeze it quickly.

He experimented with a regular refrigerator. It works by convection, that is, removing heat by a cold stream of air flowing over the food. Birdseye saw immediately that this wasn't fast enough, so he invented a machine which froze by conduction, by pressing thin pieces of the food between metal plates cooled to 25 degrees below zero. In this way, both the temperature and the freezing time could be closely controlled. Even after Birdseye discovered that quick freezing was necessary to avoid large ice crystals, there remained many problems. For example, all cell structures didn't freeze or defrost in the same way or at the same rate. So, for each kind of food Birdseye had to discover the precise temperature needed to produce the smallest crystals in the shortest time.

Once he perfected his flash freezing method, he borrowed on his life insurance to set up, in 1924, the General Seafood Company. He flash froze the catch as soon as it was hauled aboard a trawler.

His Seafood company brought him a fortune when he sold it for millions. He spent the rest of his life inventing, patenting nearly 3,000 ideas. The most successful of Clarence Birdseye's inventions, aside from frozen food, was appropriately enough, an infra-red heating lamp for thawing it.

The Mouse

THE COMPUTER MOUSE is to us nothing special, yet it revolutionized how we use computers. The end of its story is 1968, a time well before the first PC had appeared.

In a San Francisco auditorium Douglas Engelbart sat at a computer console. A huge screen showed his face, overlaid with the display of a computer. Engelbart demonstrated a new device: A clunky, square wood box, just small enough so his hand could fit around it. On top it had a single button, and underneath two wheels. Englebart's presentation of his mouse astonished the crowd.

You see in 1968 computers were not interactive things. You fed in a pile of punch cards, then waited twenty minutes for the computer to do its thing. With Engelbart's mouse the computer could be used instantaneously, becoming an extension of a human being. And that, not just the invention of the mouse, was Douglas Engelbart's great insight.

The mouse began in 1945. Engelbart, age twenty, sat in the Philippines in a bamboo hut, which housed the Red Cross Library. As he waited to be shipped home from the war, he found an article in the *Atlantic Monthly*. It detailed a way for humans to index the mass of scientific information just beginning to explode. With uncanny accuracy, the 1945 article described the World Wide Web we have today.

Because of this article Engelbart decided to find a way for humans to cope with all that new information. He sketched a system of knobs and levers so that symbols on a screen could be manipulated. He put the idea aside, but it came back full force in 1957.

As he drove to work on his first day as an electrical engineer, he made a life-changing calculation. "I was 25," he recalled, "and I figured, well, 65 is hopefully when I'll quit. So that's 40 years, 2,000 hours a year, it ends up being about five million minutes." He asked himself what exactly did he want to do with that five million minutes. He returned to his thoughts about how humans deal with masses of information and that started him on his journey to the mouse.

To him it wasn't a tool, it was a part of a grand idea he called "human augmentation." A way to make the computer an intimate part of our existence.

Although his mouse is extremely successful, he isn't entirely satisfied. He finds his vision of complete human augmentation unrealized: "The personal computer," he says, "has allowed us to work better, but we still work, for the most part, alone. Relative to what our potential is," he adds, "we can go as high as Mt. Everest, and we're only at 2,000 feet."

Counting People

IN ROMEO AND JULIET, Shakespeare asked "What's in a name?" Concluding that "a rose by any other name would smell as sweet." This may be true of a name, but it isn't true of a number.

For example, an anti-war protest doesn't smell nearly as sweet if 50,000 people march instead of a half million. At a recent demonstration in Washington DC, the Capitol Police angered the march organizers by estimating 50,000 protesters, when the organizers claimed a half million. So, who's right? The answer is "we don't know" because no one is really counting.

In the past Congress mandated that the Park Service carefully count the number of people at every event at the Mall. But in 1993 the Park Police estimated Louis Farrakhan's Million Man March at 450,000— a very impressive number—but still Farrakhan threatened to sue. Very quietly, in the next appropriations bill, Congress banned the Park Police from counting at all.

Now it would seem a simple thing to count people, but it isn't. For example, the wrong thing to do is stand in the crowd and look around. When surrounded by throngs of people, most counters will estimate a crowd of 20,000 to be two million.

Instead, you must first make a careful measurement of the area to be occupied. Second, as the rally occurs go into the crowd and measure the crowd density, that is the number of people per square foot. Typically in a subway during rush hour, it'll be about two and a half square feet per person, although crowds thin out quickly to five square feet per person. Third, use an aerial photograph to capture a snapshot of the crowd and to see how much of the area is filled. Using the photo and the crowd density, you can estimate the number of people.

The problem, though, isn't with the method, its that the organizers don't trust the counters. But there might be a way around this.

At a recent anti-war march, the San Francisco Chronicle added a new twist to crowd counting. They used crowd density and an aerial photograph to calculate a number for their news reports. But then they put the high resolution aerial photo on the web, and said to their readers: If you don't like our estimate, print out the photo and count each and every soul yourself.

But before you do this, I'd advise you to keep in mind that the actual number might be a red herring. Clark McPhail, a sociologist who is a crowd expert, warns that the number might not be as important as exactly who is attending.

He notes that the 1970s anti-war protests became effective once middle America started turning out. "It was," said McPhail, "when middle-aged, middle-class folks, and doctors and lawyers" turned out that public opinion began to change."

Google

W E LIVE, OF COURSE, in the information age. Flowing at us in an astonishing amount: We have millions of books, cable TV brings in hundreds of channels, and we have the greatest pipeline of information in history—the World Wide Web.

It now has over a billion pages, hooked together by over seven billion hyperlinks—that's one link for every person on this earth.

But information, of course, isn't knowledge. The first step in turning this glut into knowledge is to organize it. For the web there is one standout: Google.

About 150 million times a day, someone uses Google to search for information on the web. Type in some words, click, and up pops a list of pages arranged—and this is the amazing part—from most useful to least useful. How in the world does Google do this?

First, Google is no amateur operation. It's a company of 650 people who focus full time on one task: How to be the most accurate and fastest search engine in the world. In 1999 a typical search took three seconds, now it's less than a tenth of that.

To organize the web, Google uses a very clever method. Its developers realized that the web could be pictured like a road map. They thought of links from one page to another as roads, where the destinations were web pages. They surmised that the pages with the most roads leading to them were the most useful.

For example, they took all of the web pages that contained the phrase "public radio", and ranked them from the most linked to the least linked. Not surprisingly, *www.npr.org*—the home page of National Public Radio—shows up first, followed by *www.pri.org*, Public Radio Internationals home page. In addition to this main method, Google uses other tricks. For example, they also look at the placement of the text: If "public radio" is in capital letters or near the top of the page, it's more likely to be mostly about that subject, than a page where it is in small type or near the bottom. So effective is Google that its meets my personal criterion for being named a superstar technology, which is this: When a technology becomes so commonplace that we use it for amusement, it's then a superstar technology. This has happened for Google: Users have invented the game Googlewhack. To play you find two search words that when typed into Google, deliver one and only one web page. For example, the words *ellipsoidal triathlete* will turn up one and only one web page.

Myself, I really like Google because it's given me new purpose in life. When you type my last name "Hammack" I come up number two. My new goal, of course, is to become number one. So, friends, start linking.

I wrote this piece in 2003 when Google was just beginning. They've grown, of course, tremendously since then. Their method for ranking pages was more complex than what I outlined and has become even richer, grander and more secret since then! Some have called it the most valuable algorithm in the history of the world.

Adam Osborne

THIS MONTH THE high tech world lost one of its most colorful figures. Adam Osborne profoundly changed the way we use computers when he introduced the first portable computer. There was nothing like it before him, but then there was nothing before like Adam Osborne.

Born in Thailand to British parents, he grew up in southern India, then trained as an engineer in England, but failed at a career in engineering. I called him "colorful", but many thought him "loud" and "arrogant." Osborne himself reported that on his first job he "quickly became the guy everyone wanted to watch slip on a banana skin." So, he left corporate life to start his own publishing firm specializing in computers.

After writing about the industry for several years, he noticed the same gripe from his readers. He put it this way: "How are you supposed to use this machine that comes in five boxes and is wired together like spaghetti?" Seeing an opportunity, Osborne sold his publishing company and used a quarter million dollars from that sale to make compact, single-piece computers.

The Osborne I, as the computer was named, weighed twenty-four pounds—it was the size of a sewing machine—had a five inch screen, and a plastic case with an unpleasant pebbly texture. But it was much more than just the idea of a portable computer that Osborne brought to the field.

First, he pioneered bundling software with the computer. Before that each was sold separately. Who, today, would buy a computer that doesn't work immediately upon unpacking it?

His second legacy was to be the first entrepreneur to go through the boom and bust of the dot-com industry, so familiar to us today. So fast did his sales increase that the term "hypergrowth" was coined to describe it. Within a year and half from opening, his business had revenues of over 100 million dollars, making his firm the fastest growing company in history. But as fast as he rose, Osborne fell.

The myth is that Osborne announced too soon the improved, second model of his portable computer, causing sales of the first to fall off as buyers waited for the new version. But the truth is that he lost ground to the giants like IBM as they introduced their PC, thus creating a standard for all manufactures, and leaving the Osborne I computer stranded. Less than three years after he started, Osborne was bankrupt.

He once said, "When you become an entrepreneur you can go up awfully fast but you can go down just as fast. One day they're famous, the next day nobody knows who the hell they are."

I think we should remember Adam Osborne. He died this month, aged 64, in Kodiakanal, an isolated village in India far from the glitter of dot-com hype.

Potholes

POTHOLES ARE A UNIQUELY American phenomenon. Drive the highways of South Africa, Germany or France and you'll find few ruts and divots. Why potholes in America and not everywhere? The answer is the roadbed—the layer under the road.

To make a road, engineers first prepare the soil: They mix it and smooth it and then compact layers of rock on top. Next they make a mixture of rocks stuck together by asphalt. Asphalt is the gooey stuff left over from distilling crude petroleum. This rock and asphalt conglomerate is dumped onto the roadbed where a paving machine spreads it to finish the road. And then, in America, potholes form.

Small cracks in the pavement fill with water, which freezes and expands the cracks. The ice melts in the spring leaving a gap and weakening the pavement, which eventually gives way, creating a pothole.

Now, in America, we spend millions fixing potholes, either shoveling asphalt in by hand, or slapping it in using expensive machines. And because some eighty percent of these patches are gone within a year we look for quick technological fixes.

For example, I read recently of a new material for patching potholes that is "harder and more dense" than concrete and will, of course, revolutionize road repair. But the key to fixing potholes is to prevent them. The secret is to spend time preparing the roadbed.

In South Africa, which has perhaps the world's best roads, they do lots of compacting, smoothing, and mixing of the underlying soil to create an even layer. This gives the road a good foundation so that when cracks do appear, they don't easily form potholes because the well-packed ground doesn't give way. This careful approach is used in most of Europe, so why not here?

Well, a pothole is not just a technological thing, it's also a political entity. Usually we think of European nations as steeped in governmental regulation, and of the United States as a free market, but actually the opposite occurs in building roads. In the United States the government sets specifications and asks contractors to meet them. Once done with the road they have no more responsibility.

Contrast this to France where the contractors must come up with their own specifications and must guarantee their work. This means that if a pothole develops the contractor has to fix it. But in the United States, once the job is done, as long as it has met specifications, the contractor is no longer liable. So, when a pothole appears, a whole new round of bids must begin. The prescription, then, for getting rid of potholes, is to give contractors adequate funds to build a strong roadbed, and in return require them to take responsibility for their road for its lifetime.

Until this happens, though, our cars will take a beating every spring.

SARS

I HAVE SEVERAL FRIENDS AND COLLEAGUES returning from Asia right now, and some leaving for China. So, I took a careful look at the mathematics of SARS—Severe Acute Respiratory Syndrome. The mathematics used to study disease is called epidemiology. Its goal is to create a vivid picture of a disease in space and time. For example, at first SARS seems like a tiny thing. Right now, there are about 7,000 cases in the world, and about 400 people have died. Each death is tragic, but it seems small compared to a world population of billions. But here is where an epidemiologist paints a precise picture that changes this naive view.

First, they look at the kind of people who are getting ill. For SARS the young and healthy people have suddenly become critically ill. This alone causes alarm. Next, an epidemiologist measures how it spreads: SARS began in the Guangdong Province of China and has rapidly reached four continents. Another cause for alarm. And lastly, they look closely at the growth rate.

They tabulate every day the number of new cases, and then compare it to the previous day to see how quickly SARS is growing. They are watching to see whether the growth is what they call linear or exponential. Linear growth is normal constant growth. For example, a child between the ages of six to thirteen grows the same amount every year—about four inches. This is called linear growth, because it's the same increment every time. Contrast this to exponential growth, which is quite astonishing.

It's captured best in an ancient legend about the inventor of chess. His king loved the new game of chess, so he summoned the inventor and asked what reward he would like. The inventor requested that one grain of wheat be placed on the first square of the chess board, two grains on the second square, four grains on the third, and so on until all sixty-four squares were covered. The king, thinking this a modest reward, ordered his servant to fulfill the request. To his surprise, all of the grain in his kingdom couldn't fill the board. By the end, the number of grains on the last square would be a nine followed by eighteen zeros. This is the power of exponential growth.

So, if each increment of growth for the number of cases of SARS grows larger than the last—this exponential growth—then we likely have a nasty epidemic on our hands. Right now it is growing linearly, there are about 250 new cases added every day. Not that less than exponential growth means the world is safe, but it seems that this isn't the big one scientists have been watching for—like another 1918 influenza, which killed twenty million people.

Yet, I suggest you keep plotting the numbers yourself and watch carefully the growth rate of SARS.

Concorde

WITH THE CONCORDE soon to stop flying, an era of air transport has come to an end. Unknown to most people the era ending is the 1950s, and the Concorde is one of the greatest failures ever to fly.

How could the Concorde be called a failure? After all, it flies higher and faster than any other commercial jet. Yet for its entire life, the Concorde ran in the red financially, although for sure it was a technological marvel.

With the largest engines of any commercial aircraft, it took off at a whopping 500 miles an hour, reaching that speed from a dead stop in thirty seconds. The sonic boom of its engines would disturb those living around the airport, so to prevent noise, the engines were throttled down only a few seconds after take off. This quiet, after incredible noise and acceleration, often terrified passengers who thought the plane's engines had failed. All they had to wait for, though, was for the Concorde to reach the Ocean: When, far away from any population, the plane came to life.

The pilots fired the afterburners and the plane zoomed to over 1,000 miles per hour—nearly twice the speed of sound, and three times faster than a normal jet—and it soared to almost ten miles above the earth—a 737 flies at about half the height. So high was the Concorde that passengers could see the curvature of the earth.

In spite of this apparent success, the Concorde has several major limitations. Although, it was the world's fastest commercial plane, designed to jump over oceans, its fuel tanks didn't hold enough to fly over the world's largest body of water—the Pacific Ocean. In addition, it was tremendously inefficient: It used the same amount of fuel as a 747, yet held a fourth as many passengers.

But two bigger things killed the Concorde: The information revolution, and a large slow jet.

In the 1960s when the Concorde was developed, there was a need to hop around the world. Today, though, with cell phones, cheap long distance service, and express mail, there are many times where a person can just stay put, instead of hop a continent.

But most significantly, the Concorde missed the market. The Concorde was built when travel was for the elite, especially business travelers, CEOs and the like. Today travel is a mass market, where cheapness is prized over quickness. Two-thirds of travel now is for fun, not business. The Boeing company guessed this correctly. At the same time as the Concorde was developed, they built the first 747—a plane that flies slower than its predecessor the 707, and whose successors—the 757, 767, and 777—fly even slower than the 747.

Apparently the Boeing engineers, unlike the Concorde ones, had read Aesop's fable of the tortoise and the hare.

HeLa Cells

WE OWE A MAJOR STEP in the eradication of polio, and a host of other diseases, to one unsung person. I'd say hero, but this person never knew what they did. They never knew their own contribution. In 1951, a thirty-one year old woman named Henrietta Lacks lay in a segregated ward of the Johns Hopkins Hospital in Baltimore. Poor and African-American, born to tobacco pickers from Virginia, Lacks herself was the mother of four. She was dying of cervical cancer.

As the hospital's gynecologist sewed radioactive radium to her cervix in an attempt to kill her cancer, he took, without her knowing, a small sample of her tumor. He passed this on to Dr. George Gey—pronounced "guy".

Gey headed a lab at Hopkins that specialized in growing tissue samples—what we'd call human cells. For thirty years he'd been trying to grow human cells in his laboratory. If he could do this, then he could learn firsthand about human biology without experimenting on human beings directly. His greatest hope was to make and study long-living cultures of the most dreaded human diseases, to have, as someone once put it "a tumor in a test tube." The problem though, was that human cells wouldn't grow in dishes.

Guy noticed that Henrietta Lacks's reproduced in the dishes—even thrived. Gey called them immortal because they were the first human cells to live indefinitely outside the body. Their first use was to develop the vaccine that wiped out polio. Today, nearly every lab using tissue cultures uses Henrietta Lack's cells. They call them HeLa Cells—spelled H-E-L-A- after the initial letters of her first and last names. They are a standard laboratory tool for studying the effects of radiation, growing viruses, and testing medications. They've been used in Nobel Prize winning work, have flown in the space shuttle missions, and sat in nuclear test sites around the world to test for radiation. In fact, they have been cultured so often the that the cells combined weight exceeds many times that of her original body.

Within the scientific community the abbreviation "HeLa" for the cells is often used. In fact, when I searched a database of scientific papers I turned up nearly 1,000 papers with the words "HeLa cells", but only one when I type in "Henrietta Lacks", the unwitting supplier of these cells. It's time we remembered.

Laundry Machines

THE REASON WE STAND in our basements and do laundry is a complex mix of scientific facts, technological innovation, and worry about class and status. It began with a scientific understanding of germs.

In the eighteenth century frequent bathing was thought to be a threat to health, and the tub was approached with great caution. Most people held a middle ages belief that wetting the skin left one highly susceptible to disease. Toward the end of the 19TH century, science linked microorganisms and disease. This led to the belief that that dirt, and thus disease, could be lurking even where it could not be seen or smelled. This was intensified when people recognized the connection between dirt and skin diseases, for example, ringworm. So, after having avoided bathing and frequently changing clothes, people became voracious consumers of cleanliness during the nineteenth century. Cleanliness became both a health issue and a way to distinguish oneself from the laboring classes. A clean, white starched and ironed shirt came to represent these new-found values. It also preserved an important illusion for the middle class: It hid the fact that one worked for a living.

So, in the late nineteenth century and early twentieth, a series of industrial laundries rose to meet this need for cleanliness. At these laundries a person dropped off their clothes, then picked up the clean bundle a day or two later. These industrial laundries would have thrived, but the idea of germs backfired. Notions of cleanliness contributed to prejudice against people—like laundresses—judged inherently unclean because of race, culture, class or ethnicity. The middle class now feared having other people touch their clothes.

This gave impetus to personal washing machines. They would never have appeared, except for two other revolutions—one technological, the other economic.

First, a reliable electrical grid was strung across American; this provided cheap energy to run the motors of washing machines, and second, the installment plan appeared. This allowed the middle class the means to buy the machines. The new American Standard of Living was bought on the installment plan. At the outbreak of World War II, the major appliances sold by installment included radios and phonographs, refrigerators, stoves, vacuum cleaners and, of course, washing machines.

If you think washing machines too arcane a way to talk about technological progress—compared especially to wonders like jet plans, and rockets, and cell phones—keep in mind that at one point the world's two superpowers thought them vital.

Recall the 1959 "kitchen debate" between Russian Leader Nikita Khrushchev and American vice-president Richard Nixon. It revolved around which cold war power was better able to "liberate" women from domestic tasks through appliances.

Duct Tape

DUCT TAPE IS A POP ICON of the technological world—but only in America. It's sold in other countries, but no where else does it have this reputation as a universal cure-all. The reason duct tape resonates so strongly in the America psyche is due to duct tape's origins: It long been part of America's battle for freedom and liberty— and perhaps the ultimate representative of Yankee ingenuity. Recently, of course, we've heard America's Homeland Security secretary earnestly advise stock piling duct tape to defend against a biochemical attack. This use echoes the origins of duct tape: It was made to combat the menace of Adolf Hitler. During World War II, the military had great trouble with water seeping into ammunition boxes—wet bullets don't work very well. In response two inventors at Johnson and Johnson set to work on a special tape to solve this problem.

They took a surgical tape made by their company and added a water-proof layer of polyurethane sealant. Because the cloth layer was cotton and because the sealant made water bead up on the tape, soldiers dubbed it *duck* tape . It reminded them of ducks because water rolled off it like a duck. The tape became standard military issue in olive green, and it instantly became a military staple. As intended soldiers used it to seal their ammunition boxes, but they also used it to mend boots, patch holes in tents, and strap equipment to jeeps. When the GI's returned home they brought duck tape with them.

In the booming suburbs of the 1950s it filled a million household needs, including, of course, sealing duct work. Hence the new name, and also the color changed from green to silver.

It has now, of course, reached pop icon status. It's been used to make prom dresses—as part of the yearly duct tape fashion show. The

Apollo 13 astronauts used it to improvise a life saving carbon dioxide filter. And sadly, it is favored by the criminal classes: It's been used to hold open doors, and to bind bodies. And it is the focus of many jokes. Alaskans claim that a cardboard box sealed with duct tape is an "Alaskan Samsonite."

Today duct tape comes in seventeen colors, including hot pink. Some 600,000 miles of tape are sold every year—enough to go around the world seventy-five times. Yet there is one thing you should not do with it: Use it on heating ducts.

Researchers at the Lawrence Berkeley National Laboratory found that duct tape lets more energy seep out from ducts than just about any other type of tape. In fact, it releases so much energy, that California denies a tax-credit to any ducts covered with duct tape.

SCUBA

I'M ON MY WAY TO THE Caribbean for a week of SCUBA diving. Every breath that I take under water is due to Jacques Cousteau. You probably think of him as just a television showman of sorts, yet he was a real innovator in the technology of underwater exploration.

At night Jacques Cousteau had dreams of flying using his arms as wings. In his waking hours, the closest he could get to this was swimming underwater. Wanting to "fly" even deeper he first used a long air tube from the surface—but this tube offered as much peril as opportunity. One day, while diving off the coast of France, Cousteau felt the air pipe break. Luckily he quickly closed his wind pipe and shut off all air from the surface. If he'd taken a breath he would have burst his lungs.

As Cousteau descended the pressure on his body increased. For example, at thirty-three feet his lungs felt twice as much pressure around them as they did at the surface. To prevent his lungs from collapsing and killing him he had pressurized air sent through the surface pipe. When it broke, it filled with unpressurized air, and a single breath could cause his lungs to implode. To wean himself from this danger, Cousteau dreamed up what he called a "self-contained compressed-air lung" or, in his more evocative phrase, an "aqualung." He wanted to strap a compressed air cylinder to his back, and have a mouth piece that delivered air whenever he needed it.

Cousteau knew of the demand system used to supply oxygen through the masks of high-altitude pilots: Their air supply flowed only when they took a breath. So he headed to Paris to find someone to invent such a device. By luck he ran into Emile Gagnan, an expert on industrial-gas equipment. As Cousteau outlined his ideas, Emile interrupted and said, "Something like this?", handing Cousteau a small device. He explained: "It is a demand valve I have been working on to feed cooking gas automatically into the motors of automobiles." At the time there was no gas for cars and all sorts of projects were under way for using fumes from burning charcoal and natural gas. Gagnan modified his valve for Cousteau, then in June of 1943 sent him a prototype.

Cousteau went to a railway station on the French Rivera and got the wooden case expressed from Paris. "No child," Cousteau later said, "ever opened a Christmas present with more excitement than ours when we unpacked the first aqualung." He added: "If it worked, diving could be revolutionized." Indeed it was.

With his new device he was free now to fly underwater. He experimented with loops, somersaults and barrel rolls. He stood upside down on one finger and burst out laughing. I know exactly the freedom he felt: I'll soon be doing my own underwater somersaults and barrel rolls in the Caribbean.

Wind Energy

ENERGY FROM THE WIND IS renewable and pollutes very little, yet the wind supplies only about one percent of the United States electricity. Why such a small amount?

There are several reasons that wind energy hasn't been universally adopted in the United States.

First, wind energy only recently became cheap. The most important piece of machinery in turning wind into electricity is a turbine. The large blades of the windmill spin the turbine, and its motion turns wind energy into electricity. A turbine, of course, is the same thing that drives a jet. So naturally the first manufacturers of turbines for capturing wind power based their designs on jet engines. But this yielded wind turbines that were inefficient, making the cost of a kilowatt of wind energy about 40 cents in the early 1980s—many times more than fossil fuels.

Today's state of the art windmill is fifteen stories tall, with blades 200 feet or more across. They move very slowly, typically about fifteen revolutions per minute, a tenth that of older systems. New turbines are so efficient that wind energy costs about the same as coal, natural gas or nuclear.

With these advances, what's the problem now?

It's this: You have to build the wind mills where there is wind. Typical places for wind farms, as they call banks of windmills, are plains, shorelines, the tops of hills, and the narrow gaps between mountains. Places rarely near transmission lines.

The United States transmission system was designed to supply electricity to a local area, so power plants are typically built near cities. Since we build our cities where the wind doesn't blow, there are no power lines near wind farms. This calls for building costly transmission lines over unforgiving terrain.

In addition, wind power differs from fossil and nuclear fuels in a critical way: It can supply steady electricity, but not a burst of electricity. Utilities use coal- and nuclear-powered plants, in addition to peak plants that kick in when demand is greatest. Engineers are designing special batteries to supply energy when the wind dies down, but the problem hasn't been solved yet.

To find the solutions we might look to other countries. For example, Denmark gets one-third of their electricity from wind. Yet, oddly this highlights the scale of the problem in bringing wind power to the United States. Denmark is slightly smaller than Vermont and New Hampshire combined and has a population about that of Chicago. To generate their electrical energy from wind takes over 6,000 wind turbines, located off-shore.

So, wind power isn't the panacea that will save us. The most optimistic estimate I can find is from the American Wind Energy Association. They think that about six percent of America's power will be from wind in the next twenty years. Mostly likely wind power will be part of a patchwork of many energy systems that, if all goes well, will supply the energy needs of the United States.

Fiber Optics

RECENTLY AN EARLY MORNING PHONE call woke me. The caller, a regular listener, had heard me talk about how glass had changed the world. He said that I'd missed the boat: The single largest effect glass has had on the world, he said, is apparent every time you make a phone call. It's fiber optic cables. Indeed he was right.

A fiber optic cable is truly an amazing and revolutionary thing. It's a piece of glass used to guide light. Yes, just like water flows through a pipe, and electricity through wires, you can guide light with a special type of glass.

The key is to make the glass as clear as possible. If the ocean were as transparent as a fiber optic cable, you could float on the ocean's surface and examine the ocean floor to its greatest depth. So clear is this glass fiber that a light pulse can travel for sixteen miles before it dims.

Engineers use fiber optic cables to send phone calls around the globe. Just like communicating with smoke signals or the dots and dashes of Morse code, the words you say into the telephone receiver are turned into coded flashes of light that travel down the fiber.

To see the effect of this ultra-clear fiber you need only call Australia from the United States. My wife recalls calling her sister, who lived there in the 1970s. At that time you'd have to say a few words then wait for them to be transmitted, then wait for the other person to respond. In the past there was a delay because the signal bounced off a satellite. Today, thanks to fiber optics, there is none; it now sounds like talking to your next-door neighbor.

Right now phone companies are moving rapidly to revolutionize the world with these fibers. Every day installers lay enough new fiber optic cable to circle the earth three times. There will soon be enough to carry trillions of pieces of information a second—not just phone calls, but e-mail, photos, audio and video. Already there are eight cables crossing the North Atlantic Ocean, and six more crossing the Pacific. In 1997 a major transoceanic cable, called FLAG—for Fiber-Optic Link around the Globe—was laid between London and Tokyo, with branches to Spain, India and China. In fact, China is one of the most enthusiastic advocates of fiber optic cable, laying some two and a half million miles so far.

It's speed they're after: It would take hours to download a full-length movie on the current copper phone lines that come to your house, but with a fiber optic connection, it happens in seconds. By 2010 we'll have enough fiber optic capacity so that every person on the globe could download a dozen movies in a few seconds.

I suppose this is progress, but what exactly we'll do with a dozen movies a second is beyond me.

Spam

RIGHT NOW AMERICA ON-LINE filters out 780 million unsolicited e-mail advertisements, commonly called spam. To put that in perspective: That's 100 million more than they actually deliver. Yet our in-boxes still fill with spam. Why is this spam so hard to filter?

Getting a computer to filter spam automatically is difficult because spam deals with the essence of what makes us human—and different from anything else on earth—communication by words. It is the content of the message that determines whether or not we want to read it. That's why the most common filters fail.

Existing spam filters look for specific words in the message. For example, I get offers for inkjet cartridges, so I exclude the word "inkjet". But now the spammers have started spelling the word with dashes: i-n-k-j-e-t. And what if a friend writes to me about his inkjet printer? I'd miss his e-mail.

The key, then, is to assess the actual content of an e-mail—and determine whether to keep or toss out the message. There is hope of having your computer do this because of an 18TH century English Minister.

Very little is known about Reverend Thomas Bayes except that he left a single scientific paper with a revolutionary explanation of how to estimate the probability of an unknown event—an event like an e-mail arriving at your mail box that you want to read, or don't want to read.

Over 200 years later a Harvard-trained computer scientist, Paul Graham, wrote a paper called "A plan for Spam" showing how to use Bayes work to rid the world of spam.

Graham wrote a program to rate the "spamminess" of each word in an e-mail. If 90% of the spam, for example, has the word "Viagra" in it, then Viagra gets a negative rating of 90 per cent. To prevent excluding message he wants, his program rates words that turn up in regular messages too. If his friends often e-mail about going to a movie, for example, the program takes into account the words likely to be used in such a message.

So, when Graham's program sees a new e-mail, it calculates how likely it is to be spam or desired mail, based on all the words in it. Graham's methods can keep up to 99% of spam out of an in-box, with rarely a personal message lost.

This filtering is unlikely to be defeated because it assesses the content of a message. To evade the filter, spammers would need to change their message to be like one from your regular correspondents, and that would be one with no sales message in it.

I think Reverend Thomas Bayes would approve of this use of his mathematical work. He once wrote: "So far as Mathematics do not tend to make men more sober and rational thinkers, wiser and better men, they are only to be considered as an amusement, which ought not to take us off from serious business." Surely he would approve using his mathematical work so I can get fewer e-mails about Viagra.

LEO Computer

I HAVE A TALE TO TELL YOU with a moral for today's computer business. It's the true story and cautionary tale of the world's first office computer. And its source is the most unlikely entity: A British tea shop.

The British firm J. Lyons and Company sold tea. Or at least they did at first. Founded in 1887 they found a niche in feeding the crowds at exhibitions and trade fairs. By the 1940s the company led the field, but also made and sold all sorts of cakes, confectionery, coffee and tea, and ran a chain of teashops and restaurants, one of which seated 4,500 people.

The company ran on a culture of self-sufficiency doing everything in-house—from manufacturing new bread ovens to operating tea plantations. They also printed the packaging for every product, and even laundered their waitress's uniforms.

Keeping track of this huge amount of activity took a whole fleet of clerks and accountants. To head them they hired, in 1923, a star Cambridge math graduate. This man, John Simmons, set up a think tank that guided Lyons every logistical move.

By the 1940s, Simmons and his staff were dreaming of a machine that could track their accounts. They knew of ENIAC, the computer developed at the University of Pennsylvania to crunch numbers for the military.

So, just after the World War II, Simmons convinced the Lyons Board of Directors that they should use that newfangled thing called the computer. Now there were of course, no computers to buy off the shelves at the time, it was 1947. So, they formed an alliance with Cambridge University to build their own. Called the LEO—the Lyons Electronic Office—it succeeded in keeping track of all the buns, teas, and crumpets made by the Lyons Tea Company. Based on this success they chose to expand even further. They formed LEO Computers Limited to sell ... computers! Here Lyons tried to duplicate their success in the manner they did with tea. They tried to control everything: building the computers, writing the programs, making input tape readers and output printers, and they even invented their own programming language. The LEO Computer Company died a quick death as it was acquired by a company that was soon acquired by another. In the midst of all this IBM came on strong, taking over the world with its computers.

The computer world has now gone the way of specialization, so what is the caution here? I've been keeping an eye on Microsoft. A company for whom the Windows operating systems is as central as tea was to the J. Lyons Company. I've noted that Microsoft has now moved into making XBoxes for gaming, started Microsoft TV , operates the MSN web site, runs a cultural magazine called *Slate*, and even makes watches—all of which lose money. Does this sound familiar? Perhaps they should study the J. Lyons and Company tea shop-computer story.

Volkswagen Beetle

THE LAST VOLKSWAGEN BEETLE has rolled off the assembly line in Mexico. Long gone from the United States, the Beetle found a second life in Mexico. But now, with the closing of the Mexican plant, an era has ended.

Selling over 21 million in its lifetime, the Beetle achieved pop status, gaining a reputation as the alternative lifestyle car. Yet its origins were alien to this sensibility.

It began in Berlin in 1933 when Germany's new Chancellor, Adolf Hitler, claimed that a nation would in the future be judged by its miles of paved highways. Promising to build a vast network of roads, he bullied German automakers into producing a people's car—a Volkswagen—to fill his new roads.

Hitler tapped Ferdinand Porsche to design his people's car. Porsche had built luxury cars and sports cars, yet he'd always wanted to build a simple, inexpensive car. So, Porsche outlined for Hitler a car with a small engine, four-wheel independent suspension and able to go 60 miles per hour. Hitler added a few design criteria of his own, insisting the car be a four seater to accommodate a family, and that it be air-cooled to protect the car in all types of weather—garages were few and far between in those days.

You probably recall Porsche's Volkswagen Beetle as a cheap thing, yet to my engineer's eye the car was a marvel, even a revolutionary vehicle—introducing innovations adapted only years later by other cars. Its engine and transmission were a triumph of light-weight metal alloys. Its air-cooled engine could travel in all climates, unlike traditional cars with water cooling. And aerodynamically the car was the cutting edge: Today it seems bulky but at the time, 1938, it was the most aerodynamic thing on the road.

It took Porsche's revolutionary car years to make it to the road. Hitler started the car on its journey, but throttled it with his world war: The plant designed to build the Beetle instead made missile parts.

After the War the Volkswagen plant restarted and, using Porsche's design, it made automotive history. It started with a few hundred cars in 1946, expanding rapidly to huge growth in the 60s and 70s as the Beetle took America by storm.

Yet, in the end, the Beetle proved too primitive for "modern" American. It was a cheap car because there were no air bags, no sophisticated safety bumpers to weigh it down. No emission controls distracted from the elemental simplicity of Porsche's Beetle. By the 1970s regulations from the Environmental Protection Agency and others made it impossible for the basic Beetle to survive.

So, the Beetle moved to Mexico, where it thrived until recently, killed by the another government program. Since 1994 Mexican consumers have had more car choices because of the implementation of the North American Free-Trade Agreement, better known as NAFTA.

Sliced Bread

W E'VE ALL HEARD THE PHRASE "Its the best thing since sliced bread." But was sliced bread really such a great thing? Yes! Sliced bread was the culmination of a century of technological innovation.

The end of the slice bread story starts with an Iowa Jeweler and peddler named Otto F. Rohwedder, who preferred to be called "Roh." At the beginning of the 20TH century, toasters were in vogue, and so Roh often heard on his sales calls to grocers that customers liked toasted bread, but found it difficult to slice the bread so it'd fit into the slots.

The toaster represents the crest of one wave of technological innovation that brought us sliced bread. It began with a huge effort to electrify the nation. Once homes were wired this created a demand for household appliances, one of which was the toaster.

So, propelled by the demands of the toaster, Roh built his first bread slicing machine in Monmouth Illinois. By 1917 he had a crude, working model, when fire consumed his workshop destroying the machine. It took Roh a decade to finance a new model.

A baker in St. Louis, Gustav Papendick, bought the second slicer produced by Rohwedder. He improved the cutting action, but found bakers objected to sliced bread: They felt the loaf would dry out too quickly.

So, Papendick set out to invent a machine that would wrap the bread, and keep it fresh. To do this he needed to keep the sliced bread together long enough so his machine could wrap it. He first tried rubber bands and then metal pins to keep the loaves intact, but both failed. Finally a simple idea hit him: Put the bread in a collapsible cardboard tray, which would precisely align the slices so a machine could wrap them.

This sounds simple, but again a series of technological innovations had to occur: If you are going to have uniform bread trays, you need uniform bread. Here's a few things that happened to give us uniform bread.

First, an engineering genius invented an automated flour mill. No longer was it made by hand, it could now be made in vast quantities. Then the technology for making identical loaves of bread evolved. For example, the first ovens used embers, which cooked the bread slowly and unevenly. Industrial ovens were invented that blew an even stream of hot air past the dough. In addition, the ovens were tunnels where the dough went in one end and out the other so that bread could be mass produced. This produced uniform bread, which could then be sliced automatically.

The problem of the loaf drying out remained. And here another technological revolution saved the day: Plastic came about, providing the perfect moisture-proof wrapper for a loaf.

All of these innovations came together in the time of the St. Louis Baker, Gustav Papendick. His sliced bread made sales in St. Louis jump by a whopping 80%. And most importantly it gave birth to that phrase "The best thing since sliced bread."

Indeed.

Maclaren Strollers

WHAT DO JACKIE ONASSIS, Woody Allen, Mel Gibson, Uma Thurman, Ethane Hawke, Caterine Zeta-Jones and Michael Douglas all have in common? They have used the most advanced stroller in the world the "Maclaren."

The modern stroller began in the early 1960s when a retiree picked up his daughter and new granddaughter at a British airport. To help them through the airport he had to struggle with a heavy, bulky stroller. This battle marked the end of old-fashioned strollers. Strollers of the world had no idea what they'd come up against with this retiree, Mr. Owen Finlay Maclaren.

Maclaren, a retired aeronautical engineer, had worked on the Spitfire, the first all-metal fighter plane built for the Royal Air Force. Maclaren had helped design the Spitfire's undercarriage, or landing gear. He knew the essentials of successful landing gear: It must be very strong, yet also lightweight—and, of course, it must fold up into the plane. This gave Maclaren all the knowledge he needed to improve the plight of mothers and fathers around the world.

Maclaren used his knowledge of lightweight load-bearing airplane structures, to make a durable, lightweight stroller. In 1965 he used some aluminum tubing, a piece of canvas and lightweight wheels to build a prototype. His stroller weighed less than six pounds, and collapsed with a flick of the wrist. The key to Maclaren's stroller was the x-shaped mechanism underneath and at the back: these two x-shaped hinges let it get narrower as well as flatter when folded. Older, heavy strollers folded flat, but since his folded up tall and thin it was much easier for a parent to carry or store.

He filed for a patent in July of 1965, and his first stroller went on sale in 1967. That year he made 1,000 in a stable converted to manufacture the strollers. Within nine years production was over 600,000 a year, and today about ten million have been sold.

But don't think this stroller is only a relic of a World War II prop plane. Maclaren kept his eyes on the new jets of the age in designing his stroller. I want you to examine the wheels on a stroller. In his first prototypes Maclaren had used a single wheel on each axle, but found he could steer straighter if he used double wheels, and the ride improved if he put a tiny suspension spring between each pair of wheels. If you would like to see the inspiration for this, the next time you're at an airport, examine the landing gear of a jet—you'll find it nearly identical.

Nitinol

I HAVE IN MY MOUTH THE MOST marvelous thing: It isn't edible at all, in fact it's kind of boring looking. It's the archwire that threads through my braces. This slender wire, which supplies the force that moves my teeth about one millimeter a month, is truly a high tech marvel.

It withstands long-term assault in the body's harshest environment. The human mouth contains acids of all types—from food, from the digestive system. Biting and chewing is like subjecting the wire to a jackhammer. And toothpaste and brushing is highly abrasive to the wire. Yet, all that isn't what makes it special.

The wire in my mouth is made from a novel metal that has a memory. It can be bent, then when heated will return to its original shape. It began life in 1958 when William Buehler, a metallurgist at the U.S. Naval Ordinance Laboratory, searched for a metal for missile nose cones.

He wanted one that could withstand the forces and heat of re-entry. While testing various alloys, a mixture of metals, he noticed that an alloy made from the elements nickel and titanium behaved differently from others. He demonstrated its new property at a lab meeting in 1961.

His lab assistant pulled out a strip of the alloy folded like an accordion. Then a lab scientist, who was also a pipe smoker, used his lighter to heat the strip. To everyone's amazement it stretched out to its original form.

Buehler named this memory material Nitinol for nickel, T-I for titanium, and N-O-L for his work place the Naval Ordinance Laboratory.

It works because the atoms in the metal rearrange into a new phase. Usually changing phase is something dramatic: An ice cube melts changing phase from a solid to a liquid, or we put water to boil on the stove and it changes phase from a liquid to a vapor. Less well known are that such changes occur within in a solid: Solid-to-solid phase changes involve rearrangement of the position of the atoms. Nitinol remembers its shape because above a certain temperature it returns to a rigid arrangement of the atoms.

There are thousands of uses of a shape memory alloy like Nitinol. For example, in a greenhouse, when the temperature inside gets too high, hinges made from Nitinol remember their original shape, and snap open.

In the case of my teeth the Nitinol wire is pliable at room temperature. This gives the orthodontist the flexibility to maneuver the wire into place. Then at body temperature, the wire snaps back to its original shape applying a gentle force to my teeth.

Perhaps the most interesting use is a shirt woven from nylon fibers with Nitinol wires interspersed. When balled up for packing, it can be unwrinkled with a blast from a hair drier. And when the wearer's body temperature rises, the shirt's sleeves automatically get shorter. It only costs, of course, a modest four thousand dollars.

The Power Grid

A FEW MINUTES AFTER 4 P.M. Eastern time on August 14, 2003, the largest blackout in United States history hit the east coast. An 800 megawatt power surge roared from Ontario to New York City, shutting down power grids across the region. We called this a blackout, yet, the real reason our world came to a halt was not because of dimmed lights, but because of still motors. The motor is what made electricity a superstar among energies.

We don't often think about it, but electricity is best thought of as a way to move motion from one place to another. The immediate source of electricity is motion—the mechanical motor of a generator spins to create electricity. This motor is driven by various fuels—mostly coal, petroleum and nuclear—yet, turning these fuels into electricity is a costly conversion process. We do it, though, because of the convenience: Electricity is easily transmitted, and can run motors in our homes. Without electricity each house would have a loud motor churning away to run everything. In fact, our houses would look much like a factory of the late 19TH century. Factories of that time were powered by large steam engines. A steam engine spun a shaft that ran through the factory; machines hooked onto the rotating shaft with an elaborate system of belts. That all change in the early part of the 20th century when an electrical grid was laid across the United States, and when Nikola Tesla invented a durable electric motor—a motor that runs our homes today.

Our refrigerators have motors, as do hair dryers, VCRs, and air conditioners. And, of course, many of the things in our homes exist because of electric motors. During the blackout, dairy farms lost the ability to milk their cows mechanically, and no motors whirred in ATMs to spit out money. In Detroit the blackout caused a gas shortage when it stilled the electric motors at the pumps. And perhaps the most devastating was in Cleveland, which lost fresh water because no electric pumps could move the water from Lake Erie. Some areas ran the risk that their major reservoirs would run dry and fire departments would have no way to put out fires.

What were we left with then with no moving parts that could get us through a blackout? A technology almost exactly the age of the motor, in fact one which grew up slightly before it: The 100 year old technology of ham radio. Run off of batteries, they are not dependent on our modern infrastructure. In the New York area alone about 100 ham radio operators worked with the Red Cross to coordinate the emergency response of ambulances.

Cryonics

TED WILLIAMS, THE BASEBALL GREAT, has been making headlines in the last year. Or, more exactly, his body has made news.

His head is stored in a steel can filled with liquid nitrogen, and his body stands upright in a 9-foot tall cylindrical steel tank, also filled with liquid nitrogen. Ted Williams has been frozen, with the hope that at some point in the future he can be revived. What has happened to Williams is called *cryonics*, in technical terms postmortem freezing.

If you've been frozen after death, can they bring you back? The key is what happens to your cells. The cell is surrounded by a membrane that keeps it intact. As the cell cools the fats inside it solidify, breaking down the cell wall, allowing water to leak out. The water then turns into ice crystals with sharp edges that stab the cell wall. In other words, ice turns into the cellular equivalent of ground glass. As the temperature is lowered to about minus 95 degrees the major organs in the body begin to crack. As one scientist put it "Everything is crushed ... complete pulverization and destruction. There is not a single salvageable piece of anything ... inside those cells."

You would think that this would put most people off from having themselves frozen, but those interested in it argue that while the probability of being resurrected is small because of this cellular damage, it is much, much greater than the probability of being brought back to life after being buried.

Those who want to be frozen can join a group like Alcor, which calls itself a "Life Extension Foundation." For the princely sum of $120,000 the Alcor Foundation will freeze you when you die. As an Alcor client you get a bracelet with instructions for the event of your death, warning that "no autopsy or embalming" is allowed. The moment you die they give you a large dose of heparin, which prevents your blood from clotting, and a mechanical thumper beats on your chest to keep the blood circulating. On arrival at the Alcor headquarters in Arizona, your blood is replaced with anti-freeze, and you're stored in a large stainless steel container called a dewar. It's kind of a body-sized thermos bottle.

Of course, if you're revived you'll need money. So, Alcor takes part of the $120,000 fee and invests it in a trust fund for when you return to this world. This can be very lucrative: If you invested a dollar today, it would likely be worth about 77 billion trillion trillion trillion a millennium from now.

If you have a cash flow problem today and can't afford the $120,000 price tag, you might consider what you get for $50,000—for that amount Alcor will freeze just you head. I'm sure you could buy a body with part of the billion trillion trillion trillion dollars you'll have.

Hazel Bishop

THERE IS NOTHING MORE feminine than lipstick—nor as ancient, it dates to the Sumerian's in 7,000 B.C.—yet the chemistry of today's cosmetics owes their origin to what was exclusively a man's world—the World of Business.

The first modern cosmetic was smear-proof lipstick. It was invented by the chemist Hazel Bishop. Like many women of her time, Bishop found work in traditional male-dominated areas because of World War II. She worked as a chemist at Standard Oil developing a special gasoline for aircraft engines. After the war she followed advice from her mother, who told her to "Open your own business even if it's a peanut stand."

Like all good entrepreneurs Bishop drew strongly on her personal experience: She knew the trials of being a professional woman in the work place. This brought her to lipstick. She knew first hand the embarrassment of lipstick on coffee cups and cigarette butts, as well as the inconvenience of reapplying lipstick numerous times a day. It would seem that a woman should just ditch the lipstick, but in the 1940s it was the hallmark of femininity.

Bishop spent two years doing over three hundred experiments in her kitchen to develop smear-proof lipstick. She made a lipstick with staining dyes, colorants which actually stain the skin rather than simply coat it with colored wax. She began by grinding the color into oil, then adding molten wax, which she solidified into a cylindrical metal mold. As a final step, she singed the tip with a flame for about a half a second to create a smooth and glossy finish.

To promote her new lipstick she turned to an advertising agent, named Raymond Spector, famous for popularizing the Lone Ranger. In exchange for stock in the company, Spector created a 1.5 million dollar advertising campaign to launch the lipstick.

He marketed Bishop's smear-proof lipstick as "kissable." "Never again," read the ads, "need you be embarrassed by smearing friends, children, relatives, husband, sweetheart."

The ad campaign worked: Kissable lipstick raced to the top of the the charts, selling by 1953 over ten million dollars' worth of lipstick. Yet oddly, this marked the end of Hazel Bishop's association with lipstick. She and Spector had disagreed about how to run the company. Spector bought out the other shareholders to become majority owner, and then forced Hazel Bishop out of Hazel Bishop, Incorporated giving her a cash settlement of $250,000, but keeping her name.

Bishop started other companies until settling into being a highly valued stock analysts for cosmetic stocks. She also offered advice to women: "Women should use make-up," she said, "to accentuate their most attractive feature. After the age of 25 or thereabouts, personality becomes an increasingly more attractive feature."

Hazel Bishop died with plenty of personality at the age of 92.

Monocultures & viruses

L ATELY OUR NATION'S COMPUTER networks have been attacked by virus after virus. While they aren't new—the first computer virus appeared in 1979—they have increased in their ferocity and their frequency. Why so many attacks now? The reason we have so many computer viruses now is their similarity to biological viruses. In our bodies a virus replicates itself, battling our immune system. Exactly the same thing happens in a computer: The virus, which is a kind of computer program, tells the computer to make millions of copies of the virus, until the computer grinds to a halt. So occupied is the computer with this copying that no word processing, or checking of email can be done. These virus arrive at your computer in many ways. They often come through e-mail, especially as attachments. And that highlights the problem of why we have such a large spread of viruses right now.

Our computers exists in what a biologist would call a monoculture. That's where a single organism dominates an ecosystem. For example, there are managed forests grown for timber that contain the same type of tree. These monocultures are very sensitive to attack: A disease that can wipe out one tree, can attack the whole forest. In contrast, in a ecological area that contained many species of trees, this wouldn't happen, there would be no immediate host for a virus to jump to because the tree next to an infected one might be of a completely different species that is impervious to the virus.

In exactly the same way our computer culture is largely monocultural: The dominate operating system is Microsoft windows. This is like a forest of identical trees: Develop a virus that infects Windows and you have lots of hosts to pass it to. This isn't to say there is a problem with Microsoft products *per sae*: If we had a monoculture of Macs, for example, it would be prone to the same problem. As the dominance of a single operating system continues, we can expect the virus attacks to increase.

Right now computer viruses seem like a nuisance, but the reality is more chilling. So far they've forced a shutdown of the *New York Times* newsroom, stopped the computer that routes trucks for a large east coast freight company, and halted Air Canada's reservation computer network. But these seem benign compared to the most chilling example.

A computer virus infected the safety monitoring systems of the Davis-Besse nuclear power plant in Ohio. Once inside, the virus quickly spread, shutting down all of the computerized displays which monitor vital safety indicators such as reactor temperatures, radiation sensors and coolant systems. It was lucky escape, the plant was off-line because of repairs, but the implications of this computer virus infection are terrifying.

Philo Farnsworth

THE NEXT TIME YOU DRIVE by a farm, look carefully at the rows of plants, and then think of television.

That's exactly, what happened to a farm boy named Philo Farnsworth. He'd studied science magazines and developed as someone later said, "a romance with an electron." As he mowed his father's farm he thought about about electrons and magnets. One day, in 1922, Philo stopped his horses, looked over his shoulder at the mowed rows and came up with ... television! These rows gave him the idea to use lines of electrons to make a television picture. He wanted to put his idea into practice, but he faced, like all teenagers, the perennial problem of being short of cash.

So, he waited three years until luck solved this problem for him when a charity fund raiser named George Everson from California came to town. His car broke down, and when two auto mechanics couldn't fix it, in stepped Philo Farnsworth to solve the problem. This impressed Everson and the two became friends. During their many conversations Philo talked of his ideas for television. He must have been a very persuasive teenager, because Everson believed Philo could make a television system. Everson asked Philo "How much do you need?" Philo's reply: five thousand dollars. Odd as it may seem, the man was saving six thousand dollars to gamble on something; now I can't think of much longer odds than a teenager inventing TV. With this investment, the now 19-year-old farm boy married his sweetheart and moved to Los Angeles to refine TV, but only after his mother cosigned the marriage license because Philo was under the age of consent. He worked making TV sets, until at age 20, Philo struck gold. He wrote to Everson, "The Damned Thing Works." But with the idea proven, the dam broke, and in flowed many engineers who worked to make TV an industry, not just an idea in a teenager's head.

Just as Philo started young, he became a dinosaur by age 43 as television turned into a business. In 1958 he showed up on Garry Moore's television show *I've Got a Secret*. The host called him Dr. "x" and Philo said to the panel, "I invented electronic television in 1922—at the age of fourteen." His identity was not guessed and he won a check for $80.00.

The rest of his life was spent battling depression and chemical dependency. At one point he barred the word "television" from his home. He died in 1971, aged 64.

So tonight, when you watch TV, think of yourself at fourteen and then recall farm boy Philo Farnsworth and his romance with an electron.

George Eastman & Kodak

WHEN I WAS TEN YEARS OLD my mother gave me an old Kodak Brownie camera. I was disappointed because it looked like a box with a hole in it. I didn't realize how this simple box revolutionized photography; that it changed the way American families think of themselves and recall their own histories.

The Brownie camera was the brainchild of George Eastman. In 1871 this seventeen year old bank clerk took up photography. It wasn't a simple thing in those days, in Eastman's own words it took "a pack-horse" load of equipment—including a sink because making photos was messy work. It involved coating glass plates with egg whites. His first step was to get rid of the sink, to make the process dry. Eastman worked in his mother's kitchen to make dry plates, even boiling his chemicals in her tea pot. He went into business as the Eastman Dry Plate Company. Eastman felt he could make big money from his plates, but only if there existed a small, simple camera to use them. This started him on a twenty year quest.

His first camera, in 1885, introduced a key feature: A roll of film. Eastman took the coating from his dry glass plates and transferred it to flexible paper. Although it was now convenient to take pictures, it cost forty-five dollars for the camera—an exorbitant price in 1885.

Over the next three years, Eastman improved his camera, but it still cost twenty-five dollars—again too much, although it carried, for the first time, one of the greatest trademark names ever.

To name the camera Eastman looked for a simple word that could be pronounced in every language. Eastman's favorite letter was "K"—he said it was "strong, incisive", "firm and unyielding." From this feeling he conjured up *Kodak*. With profits from these cameras Eastman spent ten more years perfecting his ultimate camera—the

Brownie. It sold for one dollar plus fifteen cents for film.

In its first year—1900—five thousand of them flew off the shelves, spreading across the globe. In 1904, for example, when the Dali Lama came down from his Tibetan capital for the first time, he brought with him his Kodak camera. In spite of the success of the Brownie, Eastman continued creating new cameras until he got a painful spinal condition that made him inactive. Always the man of action Eastman made a plan.

He tidied up his will, then asked his doctor to show him exactly where his heart was. In 1932 George Eastman shot himself through his heart, leaving behind a yellow-lined piece of paper with the words: "To my friends, my work is done. Why wait?" And what work that was.

This year alone Americans will take seventy billion photos—not simply photographs, but memories to be shared for years -- all started by George Eastman and his brownie camera.

Theremin

N O DOUBT DURING THIS Halloween season you will hear some movie, or some recording that has this familiar, yet eerie sound. That sound gave birth to the greatest gift from engineers to the arts: The electronic synthesizer.

The synthesizer began in the 1920s with Professor Leon Theremin. In a Leningrad engineering lab he played around with the latest technology: radio. It fascinated Theremin because radio changed electricity into sound. He brought two parts of the radio close together so they made a sound, like the squeal from putting a microphone too near a speaker. This propelled him, in his own words, "to give [these sounds] a musical soul."

He built an instrument where instead of physically bringing the two parts together, the performer's body would create the squeal. He would just wave his hands in front of the instrument plucking music from the air.

You've likely heard the Theremin, as the instrument became known, in the 1950s Sci-Fi classic *The Day the Earth Stood Still*.

But well before that Theremin toured the world and captured headlines. The *New York Times* called it "Ether Music." The *Chicago Tribune* said that Theremin "Mysteriously Reproduces Music." Einstein called it "as significant as ... when primitive man ... produced sound from a bowstring." The instrument made quite a splash until 1938 when Theremin disappeared abruptly.

Kidnapped by Soviet agents, he was sent to a labor camp until he agreed to work for the KGB.

But Leon Theremin had planted a seed. In the late 1950s a 14 year old boy built a Theremin from plans he found in a magazine. By age 20 he began making them commercially, selling enough to pay for his engineering education. The student, Robert Moog, used what he'd learned about electronic music from the Theremin and built, in 1964, the world's first synthesizer.

With Moog's synthesizer, the child of Leon Theremin's wonderful instrument, electronic music became world famous with one of the best selling classic albums of all time: *Switched on Bach*.

The Telegraph

MY WIFE AND I JUST INSTALLED broadband internet access, so now we surf the web at tremendous speeds. Less than 10% of Americans, though, have adopted broadband. Why so few, especially when a fast connection promises—if you believe its cheerleaders—nearly everything. One proponent even claimed that it would forge "a common bond" that would "stave off .. ethnic hatred and national breakups." Perhaps the 90% who have not subscribed know something special. Maybe a fast connection won't improve the world? Just look at the telegraph, whose story is a parable for our fast internet age.

The first commercial telegraph debuted in 1846 when Samuel Morse sent the message "what hath god wrought" between Washington and Baltimore. And with that he started a revolution in fast communication. Within eight years a transatlantic cable linked the United States to England and thus to the world, allowing a message to zip from the United States to India in minutes.

Indeed the telegraph changed the world. Prior to its invention, information traveled so slowly that the Battle of New Orleans in 1815 was fought between Great Britain and America two weeks after the two nations signed a Peace Treaty. The telegraph, though, ended the age of information isolation. By the late 19TH century money moved rapidly around the globe as banks wired millions a year. Newspapers moved from a local focus to reports of events around the globe as foreign correspondents wired in stories. Markets changed as information moved rapidly. For example, fishermen telegraphed statistics about their catches to retailers who sold the fish before they even arrived.

The telegraph did change the world, yet its promise and hype were so much greater. An early promoter of the telegraph hailed it as nothing less than an instrument of world peace, calling the telegraph and the new trans-Atlantic cable "a living, fleshy bond between severed portions of the human family," that would "bind the human race in unity, peace and concord." Instead, the telegraph gave rise to new forms of crime. In France, for instance, two brokers bribed a telegraph operator to delay stock market information from Paris to Bordeaux, then used this inside information to buy and sell stocks at a profit.

The promise of the telegraph to solve the world's problem failed, reminding us that technology and its use reflects human nature, and thus will exhibit the good, the bad and the ugly.

The telegraph itself become obsolete with the introduction of the "speaking telegraph"—the telephone. Its debut was serenaded with cries that it would "ring in the efficiency and the friendliness of a truly united people." Does that sound familiar? Maybe we should keep this in mind as the era of the telephone ends, to be replaced by the internet age.

Neon

A S A GIFT MY WIFE GAVE me a neon sign. It sits in our dining room and flashes "open" when we have a party. It's garish, of course, but that's part of its appeal—in fact, it captures some essence of Americana. Yet in the early 1900s Neon signs were the height of elegance and refinement because they came from France, the undisputed arbiter of taste.

Around the turn of the 19TH century, a Frenchman, Georges Claude, searched for a way to recover oxygen from air. What really ignited demand was an arms race: The steel industries of Europe needed tons of pure oxygen to create extremely strong steel for armaments. By freezing air, Claude was able to extract pure oxygen, but since air is more than just oxygen he also recovered a rare gas called neon.

Neon gas was so expensive that, prior to Claude's work, it was used only for exotic things. For example, in 1897 a special light display was made using neon and other rare gases to celebrate Queen Victoria's Diamond Jubilee. The glow inside these neon lights was a kind of tamed lightening. The gas in the tube is made of millions of molecules, which in turn are composed of positive and negative charges. Electricity is used to pull these charges apart and, of course, these opposites attract. When they smash into each other they produce a brilliant light. For air this gives off a white or yellow color, the color of lightning; for neon an intense, clear red is produced.

So, in making his oxygen Claude also produced tons of neon as a by-product. Previously this was rare and expensive, but with his new supply of gas he could make neon lighting available cheaply. By 1910 Parisians were able to see this beautiful soft, red light illuminating several buildings. In 1912 the world's first neon advertising sign debuted at a small Paris barber shop. A year later a spectacular sign lit up the Paris sky with 3 1/2 foot white letters "CINZANO."

Neon moved to the United States when a Los Angeles auto dealer visited France and ordered two identical blue-bordered signs with the single word "PACKARD" in orange neon letters. One sign is still functioning, having outlived the Packard car.

Georges Claude made a fortune from his neon signs, but lost most of it in the 1930s with hair brained schemes to make electricity using the temperature difference between the top of the ocean and its icy depths. He almost ended his career imprisoned for life.

Upset that the French government had not recognized his technical work well enough, he became a Nazi collaborator in World War II. Although sentenced to life imprisonment, he was paroled when France's leading scientists made a plea. He lived quietly, working on another daft plan to get power, this time by drilling holes in the ground to reach hot water. He died in 1960 at aged 89.

The Electrical Chair

A MERICA PRIDES ITSELF ON ITS technological prowess, but sometimes we overreach. To take a controversial example: The electric chair.

The electric chair rose in the 19TH century from an odd mixture of technology-worship and industrial sabotage. At the time capital crimes were punished by public hangings. They were civic lessons for teaching the perils of lawbreaking. But many felt that these executions had become rowdy spectacles, which "brutalized spectators and turned them into criminals."

As these concerns rose, electrical inventions rapidly took over domestic life. Electricity was a new and glamorous technology. It was, above all, modern. This modernization tied in with the American image of itself as a nation becoming more civilized, less barbarian.

So, many felt that although capital punishment was still needed, the morality play should be tidied up—no need for long, public spectacles, instead something neat and orderly, something reflecting the technological order of the day. And that is how the electric chair enters, although what happened was anything but orderly.

Thomas Edison's light bulb created a demand for the electrification of homes. His first power stations used something called direct current through lines buried underground, but just as Edison began to lay his wires, he was confronted by a competitor named George Westinghouse, who used an AC or alternating current system. Each system had advantages—the AC was cheaper to deliver, but the DC seemed to be safer. The public became concerned about the danger when a New York lineman got caught in a web of Westinghouse wires and was electrocuted in front of thousands of onlookers. While this incident appalled, it also highlighted that death by electrocution was instantaneous. Edison saw an advantage here. Although claiming to be opposed to capital punishment, he argued that if it were to be done, it should be done with AC current—very conveniently, the technology used by his main competitor. Edison, in a losing battle with Westinghouse, wanted AC current to be closely associated with killing. So, Edison secretly funded "independent" research into the killing effects of the DC and AC current that laid the groundwork for creating an electric chair.

Death by the first electric chair, was neither swift nor painless. William Kimber, the first man to die in the electric chair took several minutes to die and was singed by the arc of the voltage, which filled the room with the stench of burning flesh. *The New York Times* captured the result with the headline "Far Worse Than Hanging."

Yet the modern aspects of electrocution reassured Americans that killing was a legitimate state function, so long as it was done gently. So paradoxically by making executions seem easier and more scientific, it made the death penalty more acceptable and less likely to be abolished as had been done in most of the Western World—the part of the world that, in the 19TH century, was less technologically advanced.

Microwave ovens

NO DOUBT YOU HAVE USED your microwave oven this week to zap left over turkey, making use of its ability to heat quickly. Yet the oven nearly failed because at first it heated too fast.

The microwave oven is well over fifty years old. It appeared in 1946, a direct descendant of World War II military technology. A company called Raytheon produced radars for the US armed forces. The heart of a radar is a vacuum tube called a magnetron. This device produces very high frequency radio waves that are absorbed by water and fat, making them rotate rapidly, thus generating heat. The engineers at Raytheon first noticed the heating effect of radar tubes during the cold Boston winters: By accident they learned that they could warm their hands using the output from a radar tube.

Raytheon's first oven in 1946 weighed in at 670 pounds, stood 62 inches tall, and was nearly two feet in depth and width. Not only did it take up floor space, it also required a team to install it: An electrician had to put in a 220 volt line, like those used by washers and dryers today; and a plumber had to install a water line to cool the oven's radar tube. These first ovens sold for over two thousand dollars -- some 17,500 in today's dollars. It was a powerful machine: You could cook a six-pound roast in two minutes and a hamburger in twenty-five seconds.

For the next 20 years, Raytheon tried to sell the oven, but it failed to take off. The real success of the oven occurred in the mid-1960s when Raytheon acquired Amana Corporation, a successful maker of consumer refrigerators. When Amana got a hold of the oven the first thing they did was to slow it down. They redesigned it so it needed less power—and no plumbers or electricians, just plug it into a standard outlet. This slower speed reflected a change in American's eating habits and social structure since the oven's debut in 1946.

The first ovens were intended to cook whole roasts and lobsters. The patent even describes cooking "thick bodies of meat." Yet a microwave oven cooks meat, or large chunks of protein, poorly. By the mid-1960s the faster pace of life, often caused by two working parents, resulted in more packaged food, usually with smaller portions of protein. So, in the new household of the early 1970s the oven thrived as a re-heater, rather than as a substitute for a conventional oven.

Amana advertised their new oven across the nation with the slogan "The Greatest Discovery since Fire." By 1978, the microwave oven had spread like fire, rising to fifth place in appliance sales, just behind refrigerators, washers, dishwashers, and air conditioners. By 1985, it had risen to the top becoming the number one best seller.

Cooking a Turkey

IN ENGINEERING PARLANCE WE are faced this Thursday with a very tricky heterogeneous, heat transfer problem. That's a complicated way to say that cooking a turkey well is tricky.

Heat transfer is an essential part of an engineer's work. You can see the results all around you: Open you car's hood and you'll see a radiator. Or, listen to your computer. The optimal position of its cooling fan is found by a heat transfer calculation. The same techniques can be applied to figure out exactly how to cook a turkey.

The problem is this: The breast and the legs cook at different rates. The breast is composed of white meat, and the legs contain dark meat, which has more muscle and connective tissue than the white meat. The goal is to cook the turkey just long enough to break apart this tissue, so that the turkey becomes succulent. That is, the tissue turns to gelatin, which gives a velvety feeling in the mouth.

To achieve this you want the breast to be cooked to 155 degrees Fahrenheit, no more than 160 or it will be too dry. Yet, you want the dark meat to be 180 degrees and above. Under 180 it's unpleasantly chewy and even has a metallic taste. There's the problem: The white breast meat cooks faster than the leg meat, which means it's done well before the dark meat ever reaches the right temperature.

So, how can you avoid, this year, having to pour gravy all over your turkey to disguise its dryness? Here are some engineering, heat transfer cooking tips. First, make sure the turkey is fully thawed. Usually it's best to do this in the refrigerator, otherwise the turkey could spoil. The second tip is very controversial: Don't stuff the turkey. It just messes up the heat transfer. By the time the stuffing is fully cooked, the turkey is overcooked by 60 degrees or so.

Third, before cooking cover the breast with an ice pack. As the rest of the turkey comes up to room temperature, the breast will be about twenty degrees cooler. This will solve our holiday heat transfer problem: it'll slow down the cooking rate of the white meat of the breast, making it cook about as fast as the dark meat in the legs.

The last suggestion is obvious, but not always followed: use a thermometer, two if possible. Pull out the turkey when the breast reaches 155 to 160 degrees, and check that at the same time the dark meat is 180 degrees.

Then stop thinking about heat transfer and enjoy your Thanksgiving meal.

These cooking tips come from food scientist Harold McGee. You might check out his book On Food and Cooking *for more details.*

Champagne

I T'S A NEW YEAR AND I'm sure that recently you reacquainted
yourself with champagne. Here is a different view of that bubbly
beverage—a technological view.

Although champagne is closely associated with France—many there
see champagne as rooted in the soil and history of the country—in
reality champagne came about because of technological innovations
across Europe.

Now the part that is very French is the location. The cold climate of
northern France makes happen accidentally the double fermentation
necessary for champagne. Certain sugars aren't destroyed in the first
fermentation, which stops because of cold weather, and which then
ferment when triggered by the warm weather of spring. This second
step produces the bubbles characteristic of champagne.

These bubbles caused the first vintners great trouble. Up to 40% of
their bottles would explode from the pressure created by the carbon
dioxide in the champagne. To solve this problem the French got
stronger bottles from, of all people, the British.

In the seventeenth century the British learned Venetian glass
blowing techniques from Italian immigrants, the finest craftsmen of
the time. At first the British made bottles just about as weak as the
French, but then a Royal Edict changed everything. British glass
furnaces where fueled by wood, which caused major deforestation in
Britain. To save what remained of Britain's woodlands, a Royal edict
of 1615 forbid burning wood in glass-making furnaces. So new
factories appeared in Britain using coal, which provided a hotter and
more reliable source of heat. The higher temperatures allowed
stronger bottles to be made, in contrast to the French, who still used
wood-burning furnaces that operated at lower temperatures and made

weaker bottles.

The next technological innovation that helped champagne reach the market was the stopper. The French used one of wood, wrapped with hemp cord and soaked in olive oil, something that had been around since Roman times. This stopper, though, wasn't strong enough to keep in the bubbling champagne. The British came to the rescue again by using cork stoppers to keep the fizz inside, although not without help from others: The British imported their cork from Spain.

So, although champagne is closely identified with France, it was a European effort with help from Italy, Spain, and especially Britain. Even today Britain is the number one importer of French champagne —28 million bottles a year. And that interest comes right from the top: Recently a Buckingham Palace tax inventory found Queen Elizabeth had, in her cellars, four million dollars worth of French champagne.

1918 Flu

WITH FLU SEASON SLOWLY ending I'm reminded of an old joke. "A camel," it goes "is a horse made by committee." I'm reminded because the same can be said about the flu vaccine.

Every year countries across the globe assemble committees of experts to decide what should go into the latest flu vaccine. It isn't a simple task. The flu virus mutates constantly and there are many variants out there, any of which could suddenly cause a worldwide pandemic. So, they must decide whether to use the previous year's formula, or bet that some new strain will dominate.

The committees must make their decisions by late winter well before influenza attacks, because once a strain shows epidemic promise, it's too late to act. It takes time to make the vaccine—some 83 million doses in the United States alone—and it takes time for the vaccine to take effect.

The committees work in the shadow of the great flu of 1918. It is still the world's deadliest pandemic, and one which our medical system still would find difficult to treat today .

At the end of World War I, this flu strain killed over 20 million people around the globe. Nearly every section of the world suffered: The United States, Central and South America, Europe, Asia, and even remote Eskimo villages. In fact, more people died from the flu than did soldiers fighting in World War I.

What was unusual about the 1918 flu was its virulence. You see, influenza is normally deadly to the very young and the very old, but in 1918 twenty-something soldiers, although apparently at the peak of health, died within days of catching the flu. This flu strain struck hard at the most delicate of human membranes, the lungs.

Lung tissue is a gossamer net where the blood exchanges gaseous waste for oxygen. Lungs are light, elastic, and soft, yet doctors were astonished to find young men with lungs so full of fluid that they could no longer float in water.

Medical science has no clear idea of what made the 1918 flu so deadly. They fear that at any time it could reappear, killing millions. If the percentages are the same as the 1918 pandemic, then about one and a half million people will die in the United States alone, compared to the 10,000 currently killed by flu in non-epidemic years. It may be even worse in our current age.

There was only one country not significantly affected by the 1918 flu —Australia. They established a strict maritime quarantine before the 1918 flu could hit their shores, which they kept up vigorously until the winter of 1919, by which time the epidemic had ended. But today, could any country react quickly enough in the age of the jet to stop a flu pandemic? I doubt it.

Accurate Throwing

As I watched the latest Lord of the Rings movie, *The Return of the King*, I marveled at the catapults used by the dark forces to attack a great walled city. It was genius of Tolkien to use these devices, because they reflect in many ways the essence of what it means to be human. That may seem odd, but the the technology of accurate throwing separates humankind from the rest of the animal kingdom.

This may sound far-fetched, but at least one historian, Alfred Crosby, claims that it is one of three traits that define humans. The others are bipedalism, walking on two feet; throwing things; and controlling fire.

You see, walking on two feet is in many ways a disadvantage. For example, we cannot out run our prey -- how many four-legged animals can you outrun? Nor, does it make us very steady on our feet —when is the last time you saw a four-legged animal fall over? Not often. To hit a rabbit, say, which is twelve feet away, requires us to make a motion within 1/100th of a second. That's some precise calculating by our brains.

So, not surprisingly throwing things became the earliest technology. we started with rocks, then moved to making spears. There a rock was sharpened and attached to the end of a wooden rod. Next, came the bow and arrow, which allowed us to combine throwing stuff with our other defining trait, controlling fire.

The Greeks, for instance, used flaming arrows. That led in turn to Greek Fire—a napalm-like substance that could be tossed in small amounts, like a grenade, or in tubs using trebuchets. There is an argument to be made that the controlled use of fire and the ability to throw is what made civilizations spread.

For instance, the Byzantines used Greek Fire with great success. They built into their ships metal statutes of ferocious lions, then inserted flexible hoses in the lions' mouths from which they pumped the Greek Fire.

The next step in throwing things was to use fire to help us throw farther. The western world acquired gun powder from the Chinese, and this, of course, increased our ability to throw weapons hundreds of miles. Over the course of thousands of years, it led to the terrible— the Nazi's V-2 missiles and the atomic bomb—and to the miraculous —landing a man on the moon

So, in sum, throwing things represents our greatest fears—perhaps missiles will some day be tossed around the globe that will wipe all of us out; and it also reflects our greatest hope as a species. Maybe these rockets and space probes that we launch into the sky will seed the galaxy with human beings, and propagate us forever—even as we blow up the globe.

Atomic Clock

A YEAR HAS GONE BY AND that means a cesium atom has oscillated some 300 trillion times. I'm talking, of course, about atomic clocks. They work like grandfather clocks, but here the pendulum that keeps the time is the oscillation of the atom. America stores its atomic clocks in a bunker in Boulder, Colorado. Inside, fifty atomic clocks click away, sending the time to Paris.

There a team of clock-watchers combine these results with about 200 other atomic clocks from around the world. Their readings are averaged together to calculate what's called Coordinated Universal Time. It's accurate to better than one second in three million years. To my human senses that's unimaginable, I mean a year goes by and I barely feel it. So, can one second in three million years be the least bit important? Well, yes.

Here's an example that happened to me, and no doubt has happened to you. I was traveling in a jet and as we approached the ground we saw nothing but clouds. In fact, the ceiling, as they call it, was 100 feet above the ground. So, the pilot had to blindly maneuver the plane to 100 feet above the runway. The pilot used the Global Positioning System, or GPS, to put the plane in exactly the right place. As the jet approached the runway, the pilot received a signal from several satellites. The exact time it took for that signal to arrive determined the jet's position. If an error of even a billionth of a second occurred, the position of the jet would be off by one foot. That's significant for the landing I was involved with, but even more so, if the jet is an F-14 Tomcat landing on an aircraft carrier. There, a foot or two is the difference between life and death.

The accuracy of atomic clocks even affects our telephone calls! To pack lots of calls on a single line, the phone company chops conversations into tiny packets, then packs these small bits very efficiently and sends them down the phone line. At the other end, a device reassembles these bits into a coherent conversation. This is done in the time it takes to say a single word. Thus requiring a very accurate system of clocks. If all this makes you feel a slave to the clock, here's some solace: The clocks are actually slaves to us. You see a year is defined by us, by people. We define it as the rotation of the earth. But our definition isn't perfect: The earth wobbles and wiggles around its axis, causing random fluctuations in the length of a year. This means that the atomic-time grid falls out of tune with our sense of seasons, with our definition of a year. And although this year is an exception, usually the atomic timekeepers must bring their cherished, ultra-precise clocks back in line by adding a few seconds.

Proven Oil Reserves

RECENTLY, I CAUGHT A SHOCKING HEADLINE: The Shell Oil Company "lost" one-fifth of their oil reserves overnight, about four billion barrels. That is, they reduced their estimate of the amount of oil in the ground. Alarmed by this news I decided to find out exactly how long our oil will last.

As an engineer I prepared to decipher complex graphs, but what I really needed was a linguist, a fortune teller, and a philosopher of human nature.

Figuring out how much longer we'll have oil appears simple: Find out how fast we use oil, find out the barrels of oil under the ground, then combine the two. This method yields an answer of twenty to forty years of oil reserves. But, of course, it isn't that simple.

In the 1920s observers calculated that by 1930 we'd run out of oil. They overestimated the rate we'd use oil, and they didn't know about oil fields in the Middle East, South America, Africa, Siberia, Alaska, or the North Sea.

So, can we just take the current rate of consumption and use the reserves reported by the oil companies to calculate how long the oil will last?—that's how I got the twenty to forty year figure. At this point a linguist would come in handy. The oil companies report "proven reserves," defined as "those quantities of oil which are known to be in place and are economically recoverable with present technologies." Note those phrases "economically recoverable" and with "present technology." Right now we recover oil from porous underground rock. A significant fraction of the oil sticks to the rocks; we recover, at best, about eighty percent, usually a lot less than that. So, when we estimate how much oil reserves should we include this oil? Perhaps a way will be found to cheaply recover this left-over oil?

The same question applies to other sources: The Canadian province of Alberta contains the Athabasca Tar Sands, which have an oil content close to current proven reserves. And in America, Colorado's oil shale also contains vast oil reserves. Now extracting the oil isn't easy, the harsh climate in Alberta freezes the tar solid, and it takes 30 tons of shale to make 1 ton of oil.

So, the answer to when the oil really runs out comes down to a question of faith about the limits of human inventiveness: Will our technological wizards develop ways to cheaply tap other sources; or, have we reached a technological limit? In the past, bets against human ingenuity usually lost, the prophets of doom have nearly always been wrong.

Perhaps, though, in this case, we should follow the old Russian proverb: Pray to God, but keep rowing toward shore. In other words, hope that those engineering wizards will come through, but with the current oil reserves we should conserve, conserve, conserve.

Partially Zero Emissions

L ISTEN CAREFULLY, I want to tell you a secret about the auto companies. It's something that they've done almost covertly and it changes the quality of the air you breathe.

At the same time as they've built bigger and bigger environmentally unfriendly SUV's, they've created a car whose exhaust is cleaner than the air outside. Yes, that's right they have a car that produces nearly zero emissions.

I don't mean that they've created a prototype car, or have one in a research lab, but one that you could go out and buy today. Ford, Toyota, Honda, BMW, Mitsubishi, and Volvo all have versions available.

It's called a PZEV, pronounced PEE-ZEV , which stands for Partially Zero Emissions Vehicle. Like most cars today, it's powered by an internal combustion engine, and it has a tailpipe that emits exhaust, but exhaust that contains a tenth of the smog-causing emissions of typical car. In fact, the exhaust is often cleaner than the air in many cities, and amazingly the PZEV cost only a few hundred bucks more than a conventional car.

Now, why hasn't this made headlines? The reason lies in this: There was no key and startling breakthrough. As an auto company vice-president explained: "It's no one thing. It's attention to a lot of details." You see, a PZEV results from using today's technology in careful and purposeful ways.

Here are some of the unexciting details, just to give you an idea. Most pollution causing emissions are released when cars warm up, their catalytic converters—the device that treats the exhaust before it leaves the car and remove pollution—don't work well when cold. So the engineers moved the catalytic converter closer to the engine, heating it up and significantly reducing pollutants.

Next, the car makes use of the computer revolution. As the engine starts it has to be fed a mixture of fuel and air in just the right combination, if not then the fuel doesn't burn completely, letting pollutants escape. A tiny computer monitors the engine and slightly changes the ratio of fuel to air at each stage of starting.

And lastly, here is a major change that helps keep pollution down, it is so unexciting, but highlights so well why the amazing PZEV makes no headlines: They changed the leaky plastic gas tank used on most cars for an air-tight steel one!

So, you can see that a PZEV car isn't flashy. Think of the headline: "Engineer does job competently; makes incremental advance." But do watch out for these incremental advances, in the past they have often changed the world.

Lego Toy Company

THE LEGO TOY COMPANY HAS HIT hard times. In the last five years they've lost millions, cut their work force by twenty percent, and are currently in a sales slump, beaten by modern toys.

Seeing Lego decline is hard for me to watch. I'm from a transitional generation that saw Lego replace the building sets made of nuts, bolts and perforated metal strips—called Erector sets in American, and Meccano elsewhere. I recall how old-fashioned that metal toy looked next to a neat, clean plastic Lego brick. And I realize now, that at the time to adults a Lego brick suggested the simple, futuristic design of Scandinavia.

From an engineering viewpoint the bricks were pure genius as a toy: They are safe, hygienic, even pleasing to the eye, and just the right size for a child to handle. The brick opened up the technological imagination: You can combine six Lego bricks in an amazingly precise one hundred, two million, nine hundred eighty-one thousand, five hundred ways. Less specifically that's just over one hundred million ways.

Even today, a Lego brick still looks to my eye like a kind of high tech toy, but that's true only to a person of my age. The play lives of children around the world have been revolutionized in the last twenty years. How could a Lego brick look the least bit futuristic in this electronic age. The captivating world of Game Boy and XBox have displaced snapping together two bricks.

At first, it looked as if Lego would thrive in this new age. Sales grew through the 80s, but it wasn't due to innovation. The brick is the same down to the type of plastic used as when introduced in 1958. During the 80s the Lego company confused growth with success. The growing sales reflected the globalization of Lego, spreading where it hadn't been before, but not increasing market share anywhere. Not surprisingly, this has come to a stop and sales have now slumped.

Lego bricks are now hard to find. Go to a store and try to find Lego bricks, instead you'll find "Lego Kits" that seem more like models. You see, Lego has partnered with movie makers to create tie-ins: Star Wars vehicles, and Harry Potter toys.

In spite of the pressures, Lego will be around a bit longer. They may not be growing right now, but still even with the hard times, a Lego set is sold somewhere in the world every seven seconds and every year they manufacture twenty billion pieces.

Black Box

MOST PEOPLE KNOW THAT the black boxes used in airliners to record cockpit conversations are actually orange, but what is less well known is that, in a way, they were inspired by Big Band Swing music.

The black box's inventor, David Warren, worked at Australia's Aeronautical Research Laboratory in Melbourne in the 1950s. He developed special jet fuels, so he kept a careful eye on aviation around the globe. Making headlines at the time was the world's first commercial jetliner: The British Comet.

The Comet's revolutionary jet engines grabbed attention: They made the Comet's ride smooth, quiet and quick. Soon, though, the Comets made news by exploding in mid-air. David Warren had a very personal resonance with these disasters: In 1926 his father died in one of Australia's first commercial crashes.

Thirty years later, at the time of the Comet crashes, David Warren was fascinated by American Swing music, especially the big bands of Benny Goodman and Tommy Dorsey. He even used early tape recorders to make copies of his favorite songs for friends.

In the 1950s as Warren browsed an electronics trade show looking for new tape recorders for his hobby, he kept in the back of his mind the Comet crashes. At the trade show he saw the first pocket-sized recorder: A German Minifone, which weighed about 3 pounds and was just a bit larger than a hand. It occurred to Warren that if one of these had been running in the cockpit of a Comet, and if it had been recovered, it might give clues about what the pilot knew at the moment of disaster.

With that insight, he spent the next five years building and perfecting his own special recorder—the first Cockpit Flight Recorder

—only to find very little interest. The Royal Australian Air Force said that such a device would "yield more expletives than explanations." And the Federation of Australian Air Pilots declared that "no plane would take off in Australia with Big Brother listening."

Not surprisingly it was the British, the owners of the failing Comet Jets, who championed the idea. They invited Warren to England to demonstrate his device, which they named the "Black Box" because of the dark bakelite coating used to protect it in a crash. Nowadays, of course, we moved beyond this early plastic and now make the "black" boxes bright orange. With British support black boxes, manufactured in America, became standard issue on all commercial aircraft. And they are now coming to every vehicle.

Recently auto makers have installed black boxes on car, which automatically record a car's speed and other information. Today, as many as 40 million vehicles have electronic data recorders. Just like for aircraft, safety researchers, insurers and prosecutors use them to reconstruct what happened in the seconds before an auto accident.

Pompeii the Novel

I'VE NEVER COME TO YOU as any kind of literary critic. Yet today I'm going to play book reviewer and enthusiastically recommend to you a current bestseller.

I recommend to you a book, titled *Pompeii*, a historical novel by Robert Harris. Set in 79 A.D. this novel of the Roman city promises fireworks and drama because in that year Mount Vesuvius erupted cataclysmically, although the author makes the main mystery in this thriller technological.

It's hero is one Marcus Attilus Primus, simply called by his middle name Attilus. It pleases me greatly to report that this hero is one of my own ilk: He's an engineer. Yet, I hesitate to tell you the main plot because you'll think this exciting novel dull, but here goes.

Robert Harris' novel *Pompeii* focuses on a great achievement: The hydraulic engineering of the Roman aqueducts. You see, this young engineer Attilus builds a great aqueduct to bring water from the slopes of Vesuvius sixty miles to Pompeii and other towns on the Bay of Naples. Now this Roman hero finds the sound of water rushing from the mountain to Pompeii to be "the music of civilization." The mystery begins when his beloved water stops flowing: public fountains mysteriously stop, and springs no longer spout, and crisis hits the cities without water. Attilus becomes a scientific sleuth, in the style of Sherlock Holmes, as he tries to make sense of the failure of his aqueduct. Unknown to him, the problems are caused by the seismic activity of Vesuvius' impending eruption.

The author has done an incredible job of learning about Roman Civil engineering and waterworks, although make no mistake, this always reads like a thriller. We learn about the design of the aqueduct: For every 100 yards Attilus makes the aqueduct drop two inches so the water will flow rapidly. We learn that the Romans named their aqueducts—in this story called Augusta—rather than used them as anonymous sources of water. They thought that the water of each aqueduct had special properties. And we learn about the Roman discovery of cement that hardens underwater, and also exactly what kind of fish by-product the Romans used for ketchup.

I assure you, though, that it's all interesting because as the story moves forward, you know that Vesuvius will cover Pompeii with lava, ash and eventually poisonous gases. But the real treat is following the trials and tribulations of the engineer hero as he investigates, escapes danger, stumbles on nefarious plots and corruption, risks his life, and even falls in love.

And, lastly, after you read the novel *Pompeii*, you'll never again look at the water from your own faucet in quite the same way.

TiVo

MORE PEOPLE REPLAYED Janet Jackson's "wardrobe malfunction" during the Super Bowl than any other event in television history. The event, though, also gave us a glimpse of how technology might just dramatically alter television, a business that hasn't changed in sixty years.

TV operates with the same financial model it did in the 1940s. The Networks produce shows that they hope will attract millions of viewers, and the advertisers hope that a fraction of those who watch will buy their wares. They hedge their bets a bit by guessing, for example, that anyone who watches the Super Bowl is likely to be a beer drinker, and so they advertise on that show hoping to grab the attention of the few who really want to know about beer. In our digital age this is all pretty crude.

Many who watched the now infamous halftime Super Bowl show used something called TiVo. A TiVo consists of black box attached to your TV set, about the size of a VCR. It doesn't use video tape, but records onto a giant hard disk like that in your computer. At first it seems like an improved VCR, but it's potentially much, much more.

Because it's digital the recorded programs can be viewed in any order, unlike on a video tape. With a TiVo you can also put a show on pause when watching it live, and return to it without missing a thing. And if you start watching 15 minutes into a program you can skip through all the commercials. This makes it sound like a souped up VCR, but there is an additional cable out of that TiVo black box that really promises change: A phone line. Every action that you take with a TiVo is relayed to the company's computer.

There lies the promise to revolutionize TV. TiVo's computer stores a viewer's every whim, allowing the company to know the viewer better than he or she knows themself. So, it could suggest programs, including those from the past because with TiVo every show ever broadcast competes against every other show. In theory a viewer could create his own private television channel all stored on his TiVo.

The potential changes go even deeper: Because it monitor's viewing habits a TiVo box can determine when a comedian's monologue died with an audience, prompting them to tune out, or when a drama got too dull for a viewer. This offers the promise that script writers and producers can fine tune their shows for maximum appeal. And advertisers might be able to narrow-cast commercials made specifically for a particular type of viewer.

All this is one big if: TiVo must first succeed. The company has only about a million subscribers out of the more than 100 million television homes in American, plus they have yet to even make a profit.

Voting & Paper

WITH THE U.S. PRIMARY SEASON nearly over, and the general election ahead, it's time to focus on how we vote. Wanting no repeat of the Florida voting fiasco of the 2000 Presidential Election, many states now use electronic voting—just touching a computer screen casts a vote. I'm wary, though, that this new technology may simply sweep any problems we've had into an electronic void without solving them.

We are tossing out the Votomatic Ballot Tally System, invented in 1963. With these machines citizens punched holes in special index cards to cast votes, which could then be counted quickly by a computer. Most importantly, though, the Votomatic System left an audit trail—that is, actual ballots that can be recounted by machine, or if desired by hand. As we learned in Florida in 2000 what exactly constitutes a valid vote isn't clear: Does the card need to be punched completely, or need a voter only indent the card?

To avoid making these decisions, state officials are turning toward electronic voting. They hype electronic voting emphasizing how it can help voters: For example, voters often invalidate their punch card ballots by choosing too many candidates, an electronic system can alert a voter about an improper ballot.

In spite of many advantages, I'm leery of electronic voting. I've written many computer programs in my life and have learned that no software is foolproof. And I worry about security: Recently someone stole the Diebold company's voting machine software, displaying it on a web site, giving hackers a blueprint for attacking the machines.

Perhaps all this can be solved giving us nice and clean electronic balloting: Just press a button on a touch screen and the machine records your vote. Or does it?

So far, electronic voting machines have actually subtracted votes instead of adding them. And in Florida electronic voting machines found that 137 people who'd showed up and gone into the booth had no recorded votes. Did these people just choose not to vote once they got in the booth, or did the machines fail? We'll never know because electronic voting machine don't leave an audit trail.

So, the key to reliable electronic voting is old-fashioned: It's paper. The machine should record the vote electronically, but also spit out a small card with the vote printed on it. A card that can be recounted if necessary. While it seems a simple solution, many voting machine companies oppose it. It's too complicated, expensive and complex they say, arguing that it's inefficient.

When we heard cries like this—that we need electronic voting because it is clean, neat, and efficient—we should remember that democracy is often a very messy thing. In fact, there's a name for a society with efficient and time-saving voting procedures: It's called a dictatorship!

Saul Griffith

WHEN WE THINK OF THE latest technological innovation we usually think of how it affects the industrial western world—but often it also affects all parts of the world. Here's a story about its impact on the world's poor.

One day a young engineer named Saul Griffith tried to sell the Minister of Education for Kenya a new "electronic book" that he'd invented—a book that stored a complete library in it. The Kenyan minister told Saul that at least a quarter of his people couldn't even read the book because they had no glasses.

This sparked Griffith to tackle the problem of making glasses cheaply. He learned that manufacturing glasses calls for special molds to make the lenses and a laboratory in which to make them—plus a doctor to determine the prescription. In a nation like Kenya, or any rural area, maintaining molds for the thousands of necessary lenses is costly.

Undaunted, Griffith set out to automate the prescribing and manufacturing of a pair of glasses. He aimed for a cost of five dollars a pair since 80% of the families in the world earn at least a dollar a day. That way glasses would cost only a few days wages.

He designed a pair of goggles with an electronic sensor that monitors the lens in the wearer's eye and adjusts the goggles's lenses to correct the vision. This simple tool gave Griffith the correct prescription. What Griffith then did with that prescription was truly revolutionary: He built a machine to make the lens on the spot.

In the last decade or so, engineers have developed something called Rapid Prototyping Machines. These machines can make a three-dimensional object by printing layers of a thin plastic to build up a real object—not a photo, but the real thing that you can hold in your hand.

So, Griffith took the glasses prescription and downloaded it to one of these machines, which are about the size of desktop printer. It uses thin film, kind of like plastic wrap, to make a lens mold, then injects it with hard plastic to make a lens. In about five or ten minutes, out pops a complete and correct lens.

This kind of technology promises a revolution in our homes. Already through the use of computers, we now "make" many things at home that before we just could not. For example, we can "make" pictures from our digital cameras, and even "make" movies with a digital camcorder and a DVD burner. Now the next revolution of making things is just about to hit.

Perhaps in the near future, instead of waiting for a replacement part for, say, a broken washer, we'll just download some info from the internet and have our three-dimensional printer make the part.

Eulogy for Old Technologies

I T SEEMS LIKE WE LIVE in a time where things become obsolete quickly, but for every eight-track tape that has gone by the way side, there are age-old technologies that keep on giving. Today I celebrate the geriatric set of technology.

You'd think the typewriter would be dead in this computer age. Yet in 2002, Americans bought nearly a half million electronic typewriters. Even manual machines hold their own, two companies—Olympia and Olivetti—still make them. Typewriters survive because of forms: The computer word processor may be a dandy tool, but a printed form sends it scampering to the corner, while undaunted a typewriter conquers the form.

Next consider vacuum tubes. Any one over fifty can recall when slim, slick microchips replaced clunky, slow vacuum tubes. Surely in this age of miniaturization, where cell phones get smaller by the second, we can write the obituary for the vacuum tube. Yet, hundreds of millions of homes around the world all have one: The microwave oven sitting on the counter is essentially one big vacuum tube. Turns out that when you need power you cannot use a weaselly tiny silicon chip, you need the "umph" from a big old vacuum tube. And beyond that, music lovers have sustained vacuum tube technology. Many find the sound from old-fashioned vacuum tube amplifiers more pleasing to the ear.

Here's a third thing you'd expect to be disappearing: Pagers. Cell phones have become so sophisticated and so small that who needs a pager today? Yet in 2002 the old-fashioned pager surprised everyone by selling in greater numbers than the year before. Most people find a pager more reliable than a cell phone. They need far fewer transmitters than cell phones, so they provide better coverage, working in the dead spots between mobile phone cells. And pagers tend not to jam up in emergencies the way overloaded mobile phone cells do. So, many institutions still rely heavily on them: police departments, hospitals, and emergency workers.

And my favorite technology, whose death has been predicted again and again: Radio. In the 1940s many expected television to deal it a death blow. But it simply reinvented itself. Radio became the mobile medium. Cars combined with suburbs, superhighways, and longer commutes gave radio a vast captive audience. And radio also stayed local: TV covered the world, but radio gave you the weather above your head. Perhaps the age of broadcast radio is nearly over: There are now mobile MP3 players that download from the web, and satellites that deliver radio. But be wary: You'd not be the first in the last sixty years to write a premature obituary for radio.

For More Information see Ten Technologies That Refuse to Die From typewriters to vacuum tubes, these 10 technologies aren't as obsolete as you might think. *By Eric Scigliano February 2004* Technology Review

The Calendar

THIS WEEK REPRESENTS a great triumph of scientific learning. Right now we're between the Jewish Passover Feast and the Christian Easter. To develop a calendar that placed these holidays at the same season every year took the work of the best minds in science over 1500 years to solve.

The solar year, the time it takes for the earth to go around the sun, takes a precise 365.2447 days to complete. The position of the earth relative to the sun determines the seasons, but because of the uneven number of days in the solar year the seasons shifted in early calendars. For example, the weather we associate with February slowly crept into March, spelling disaster for humanity. A farmer, for instance, needed to know when to plant, or the likely date of the first frost.

The Roman Emperor Julius Caesar tried to conquer this problem with a new calendar. He declared the length of a year to be 365 days plus exactly one quarter day. In other words, he rounded that pesky 0.2447 to 0.25. His new calendar, called the Julian Calendar, used three 365-day years, plus a 366-day year every fourth year, a kind of leap year.

To get the whole Roman world up to par with his new calendar, and to put the world back on track so March would come when March should, he made the year 46 B.C. forty-five days long. He called it "Ultimus Annus Confusionis"—the year of confusion.

But Caesar's clock ran too slow by about eleven minutes, so, over a century or so, spring, again, slowly moved into winter.

The next calendar innovation occurred because of religion. The Vernal Equinox—the date when the hours of daylight equal those of darkness—determined the dates of Easter and Passover. The Jewish authorities used a lunar calendar to determine the date of the Vernal

Equinox, but the Catholic Pope didn't want to depend on another religion to determine the Christian Easter.

So in the late 16TH century Pope Gregory assembled a council to survey the best scientific work of the time and of the previous centuries, including Copernicus's earth shattering observations about the motion of the planets. The Pope charged his council with finding the exact length of the solar year, and then matching a calendar to it. They came up with a calendar based on a year only twenty-six seconds short of the true length of a solar year. They invented a calendar with unequal months and the occasional leap year—the calendar we use today.

In many ways our science today descends from this calendar because the search for a new calendar helped keep alive mathematics and astronomy in a time less than ideal for scientific inquiry. Kept alive not, though, by a love of learning, but to solve an administrative problem of the Pope.

Escapement

I N THE FALL WE performed the ritual of setting our clocks back for daylight savings time. In doing so we pay homage to the most world-changing invention of the western world: A small device called an escapement—the device that gave the clock its tick-tock sound.

You see, all mechanical clocks are driven by a weight, like a pendulum, or a tightly wound spring. The escapement restrains the motion of that weight or the unraveling of the spring, making them move precisely and evenly. The escapement made time portable, divorcing it from the natural cycles of the Earth. In fact, the first clock makers found that moving away from telling time using nature was the key to inventing the escapement.

They modeled their first clocks on nature. The sun marked the time by moving slowly through the sky. So, early time keepers built sundials to chart the sun's progress minute-by-minute. Sundials, though, didn't work at night, and couldn't be easily moved.

The next model for time measurement was the flow of a river. The Chinese built elaborate clocks where water flowed from one vessel to another. By watching the level rise the Chinese measured the passage of time. Although an elegant way to measure time, the water limited the clock to warm climates—and still it wasn't portable.

Although ingenious, these solutions to measure time erred in emphasizing a continuous movement or flow to mark the time. The essential insight that allowed time to go portable, to be completely independent of nature and its elements, was not a smooth continuous motion, but regular, precise pulses.

In the 13TH century some anonymous genius figured out how to turn the continuous unwinding of a spring into regulated bursts that moved the hands of a clock. The heart of a clock is a toothed gear, which is spun by a spring. Unimpeded, the spring would naturally turn the wheel faster and faster until the spring completely unwound. To prevent this, a small lever called the escapement, rests on the notched gear, alternately letting the gear move, then restraining it. Think of it as a seesaw sitting on the gear: It rocks one way letting the wheel move forward, but then totters down on the other side stopping the movement, repeating this until the clock winds down. This, of course, makes the clock go tick-tock.

The endless series of uniform motions from that escapement let us capture time, causing an explosion in time keeping and a revolution in our way of life: By the 18TH century the Western world was producing 400,000 mechanical clocks a year. These clocks allowed our lives to have shape and form independent of the sun and the moon and the movement of the planets, thus revolutionizing our lives. As one historian has said, "The clock, not the steam engine, is the key machine of the modern industrial age."

The Electronic Paper Trail of Terrorism

IT SEEMS IN RECENT DAYS that the terrorists have turned our high tech world against us—they've used sophisticated jets to destroy the World Trade Center, and set off bombs in Spain using cell phones. Yet by being part of the technological world of the West, terrorists organizations like al Qaeda leave themselves vulnerable to technological detection.

You see, when terrorists operate in the industrial world they leave a trail from telephone calls, e-mail messages, and financial transactions. In fact, all told, some 2,500 cables arrive every day from CIA stations around the globe, plus some 17,000 new bits of intelligence from other sources every week. This would be enough information to fill about 1,000 bound copies of the *Encyclopedia Britannica*.

That amount of information would take a human mind days, even months, to process and digest. So, today those fighting terrorism use computers to do what's called "data-mining." The military and intelligence agencies use computer programs developed by domestic law enforcement agencies to track serial killers, arsonists, and the bank accounts of white collar criminals.

The terrorist hunters feed the electronic "paper trail" left by al Qaeda into their computers, which performs something called "link analysis." The computer connects all of the dots, if you will, between apparently disparate fragments of information. A transaction at a bank in Paris, say, might appear unrelated to the purchase of an airline ticket in Madrid, but the computer ties together all these individual transactions producing a byzantine map of activity, which, if the authorities are lucky, shows the epi-center of terrorist activity.

This might seem a simple task to make the links, but the problem is branching: If one transaction ties seven people together, and these seven in turn connect with seven more, then the numbers grow very quickly. In just seven transactions of this type there can be nearly a million links to study. Small wonder that the CIA's wall's are covered with printouts as large as bed sheets laying out al Qaeda's far-flung activity. Although we like the idea of tracking silently a terrorist organization like al Qaeda, in the end we're all losers as these computer programs become more sophisticated. We like them because it means law enforcement has moved in the direction of anticipating and forestalling crime, but ultimately that requires tracking every citizen throughout his or her life—geographically, commercially, and biologically. That prospect offers the potential for enormous abuse. Although we might relish being able to silently track a notorious group like al Qaeda, we must keep in mind that the most insidious technique is the one which makes itself felt the least, and which represents the least burden, yet lets every citizen be thoroughly known to the State.

Fahrenheit 451

M Y TOWN HAS JUST FORMED a city-wide book club. By voting we choose our first book: Ray Bradbury's classic *Fahrenheit 451*. How appropriate for this age of Napster, downloading music from the web, DVDs, and electronic books. It seems at first that *Fahrenheit 451* has nothing to do with these things, after all none existed in the late 1940s when Bradbury started writing this novel, yet the the novel carries an important message for our electronic age. In writing *Fahrenheit 451*, Bradbury reacted to an event fresh in public memory: The mass book burnings by Nazis only a decade earlier. Just as the Nazis attempted to wipe out a cultural heritage, the anonymous State in *Fahrenheit 451* tries to control the past.

The book's hero, Guy Montag, works as a kind of "inverted" fireman using a kerosene-filled hose to burn books, instead of putting out fires. Montag's government has banned books, thus denying artistic freedom to build on the traditions, insights and errors of the past. We follow Montag as he converts from implementing the Government's media book ban to preserving his nation's cultural heritage. He memorizes books to keep alive the intellectual traditions of his society's ancestors.

The power of Bradbury's book today lies in its aptness for our high tech information age. Today we run the risk that every embodiment of thought or imagination may be subjected to some kind of commercial control.

It appears most often in restrictions on copying digitally stored music and movies. For example, the movie industry recently fought a court battle in California to outlaw software that makes backup DVDs for home use. This was only one of many attempts by the Motion Picture Association to get a judicial or congressional ban on all copying. We should be alarmed by these efforts, and should worry about controls on all electronic forms of information.

Think, for a moment, about electronic books: In the current climate readers may lose the rights they've had since Gutenberg's time because the publishers of an electronic book can specify whether you can read the book all at once, or only in parts. And they can decide whether you read it once or a hundred times.

Just as the book-burning firemen in Ray Bradbury's *Fahrenheit 451* nearly erase the heritage of a culture, we run the risk that the literary and intellectual canon of the coming century may be locked into a digital vault accessible only to a few, resulting in a lack of access that prevents the next generation of artist from drawing on the insights and errors of the past.

Boeing vs. Airbus

THIS MONTH THE BOEING COMPANY fired a high tech bullet at its main competitor, Airbus, by announcing the first order for its newest plane: The 7E7 Dreamliner, later named the 787.

A new airliner seems an exciting thing, yet Boeing and Airbus fight to create a plane that meets the needs of accountants: They aim for a jet with low operating costs because any savings helps in this era of razor-thin profit margins for airlines.

So, that "E" in 7E7 stands for efficiency. Boeing claims their new plane will be more fuel efficient by 15 to 20 percent over today's jets— that translates into a 10% savings in fuel cost. A big deal since a major airline spends about three billion a year on fuel. What kind of snazzy engineering creates a more fuel efficient plane? The main source of savings comes from making the airliner lighter. With the 7E7 the metallic gleam of an aluminum aircraft wing will be no more: Its wing will be made of carbon-fibers, a type of graphite, like the lead in a pencil. Engineers can make a wing as strong and rigid as aluminum, but some fifty percent lighter.

The key, though, to a competitive airliner lies not only in using the latest technology, but also in keeping it away from competitors! Airbus would love to retrofit their older planes with the latest high tech wonder from Boeing's newest jet, so Boeing does their best to keep Airbus at bay.

For example, when engineers designed the triple seven—the last new plane developed by Boeing—they set the wings high above the ground. This allowed the innovative engines developed for the triple seven to just fit, but made them too large to use under the wing of any Airbus jet. It looks like the same will happen with the 7E7 engines: They'll operate in a way that requires extensive retrofitting on any

Airbus. In spite of all this technological posturing, the 7E7 is really a multi-billion dollar gamble—a dispute, if you will, between Boeing and Airbus about exactly how air travel will evolve. You see, both companies have new planes on the horizon. Airbus is pushing its A380, which seats 555 people, making it by far the largest plane ever, outstripping Boeing's 747 by about 150 passengers. With the 7E7 Boeing moves in the opposite direction: It'll seat only two to three hundred. Boeing bets that air routes will fragment with more point-to-point flights, whereas Airbus thinks the key is a hub where thousands of passengers gather and then are rerouted.

Who's right? Well, once the giant Airbus appears followed by the nimble 7E7 we'll know by who has placed the right 4.5 billion dollar bet.

Risk

A S AN ENGINEER I'M OFTEN surprised at how people assess risk in the world, particularly, of course, the risks associated with technology.

Anyone who regularly listens to the news must think we live in a land of viperous gadgets that attack us and pollute our world. Yet, citizens of the industrialized West live in one of the safest and likely cleanest times in the history of the world. Life expectancy is up, infant mortality down, and diseases that killed millions are now eradicated.

We fear, instead, things like flying in airplanes. Yet to put that in perspective, dying or being injured in a commercial aircraft is about eight times less likely than being injured by some random object falling on us—something that I'm sure none of us spend a moment worrying about when we start our day. So, at first I'm always puzzled by this fear of technology, that I realize we fear because we're human. Our rational side, the side that deals with numbers and scientific theories, is pretty new when compared to the primitive side of our brains, which has been surviving for a long time by making choices based on instinct alone.

For example, if someone asked you to store nuclear waste in your home your gut reaction would likely be "no!" Yet, it turns out that a typical family of four would accumulate about five pounds of nuclear waste in their life time. If stored inside a thick metal case, capable of withstanding a house fire or a flood, the waste would form an object about the size of a small orange, which when placed in a thick-walled cubicle would ensure safety for you and your family. Would you have this in your home? Likely no. I wouldn't. Yet it is in some sense rational to do so.

So, when assessing risk we're not completely rational, and likely never will be. For example, I'm a bit uncomfortable flying yet I'll do a very dangerous thing without even thinking about it: I ride my bike home at night, without a light, in a dark blue business suit.

My irrational behavior highlights what I think really worries people about technology. We worry not as much about the technology as about how humans interact with it.

For example, last year at a German nuclear power plant a monitor detected abnormally high radiation levels in an employee. This led officials to question the worker, who led them to an abandoned French military airfield in southern Germany. There in a blackberry bush lay a two inch long tube, wrapped in a rubber glove. Seeping out of it was a brown solution containing plutonium. The man's motive for taking this plutonium remains unclear. This episode highlights that it's people we fear, not technology.

Douglas Adams, the science fiction writer, captured this fear when he said: "A common mistake that people make when trying to design something completely foolproof is to underestimate the ingenuity of complete fools."

eBay

ALERT EBAY USERS COULD BID last month for the tunnel boring machines used to burrow under the English Channel to create the so-called "Chunnel." Lucky bidders brought home a machine whose blades spit a half a million cubic meters of soil out its back side—a high tech machine, no doubt, yet the origins of this machine lie in a lowly worm!

These tunnel boring machines are the legacy of a French engineer, Marc Isambard Brunel. In the 19TH century the problems of digging under the Thames River fascinated Brunel. To solve the problems in digging such a tunnel Brunel took his inspiration from the greatest designer of all time: Nature.

He studied the tunneling of the ship worm Teredo Navalis—a pest that ate the wooden hulls of ships. He noticed the tough shell on the end of the worm, which it used to cut through wood. And he also noted that the rest of the worm was a long tube used to dispose of the wood shavings. Brunel conceived of a "tunnel shield" that turned miners into a huge human worm, digging under the Thames.

This shield prevented the tunnel from collapsing as the miners dug. It was a 120 ton cast iron structure, twenty-two feet tall, nine feet wide and divided into nine areas—it looked a lot like an iron tic-tac-toe diagram, but on each of the squares Brunel attached iron sides three feet deep. A miner stood in each of these opening, which were closed with fourteen three inch thick boards. The miner removed a board, dug four and a half inches into the soil, replaced the board, then removed the board below and dug four and a half inches again. When the miners were finished, workers standing behind the shield turned huge screws and drove it forward four and one half inches. As the miners began digging, bricklayers covered the newly exposed earth, building a permanent tunnel. Brunel hoped his human worm would burrow three feet a day, but he had to settle for one foot a day and it took eighteen years to build the Thames tunnel.

When the tunnel opened in 1843 it became London's biggest tourist attraction. People paid a penny apiece to walk through the tunnel which was filled with exhibitions by painters, tightrope walkers, puppeteers and magicians. Ultimately the tunnel failed economically and the bankrupt company sold the tunnel to the London railway; it is still used today as part of London's underground system.

Although the tunnel failed, Brunel's tunnel shield succeeded. His "human worm" is still used to burrow underneath rivers and lakes—including the tunnel boring machine used to create the Chunnel. Although I don't think there were any puppets or magicians ever in the Chunnel, just millions of speeding Britons.

Blackout Reading

IN THE SUMMER I ENJOY sitting in the sun reading a good story filled with mystery, heros, villains, and perhaps a catastrophe that imperils the world.

This weekend I sat on my back porch reading such a story: I read the suspense filled, although dully titled, *Final Report on the August 14, 2003 Blackout in the United States and Canada.*

It features a mighty hero brought to its knees. Indeed, the report's hero, the electrical grid, represents more than one trillion US dollars in assets, has over 200,000 miles of transmission lines, and serves nearly 240 million people. Yet this colossus was brought down in a flash.

The report builds suspense and creates dread with sentences like: "As the temperature increased from 78 degree Fahrenheit on August 11 to 87 degrees on August 14, peak load" on the electrical grid increased. I pictured vividly millions of air conditioners churning away in Cleveland, Ohio, trying to overcome the persistent heat as our hero works to keep up with demand. To no avail: At four fifteen in the afternoon on August 14, 2003 the electrical grid shutdown, stopped in it tracks like Samson without his hair.

How could this happen? The authors explain that the power grid requires the almost superhuman ability to predict the behavior of humans. You see, there is no way to store electricity cheaply in the vast quantities in which America uses it. Electricity flows at close to the speed of light, thus it must be produced at the instant it is used. And therein lies our hero's Achilles heel.

Power plants distributed across the nation must feed the grid at just the right time. Picture the grid like a balanced tightrope walker: The air conditioners pull on one side attempting to tip it off balance, but the power plants push back to keep it in balance. But if something fails then the grid can became unstable. Just as a tightrope walker loses balance and sways back and forth in larger and larger arcs until he or she falls, the electrical power grid can become unstable. On August 14TH the flow of electricity rapidly increased and decreased burning out equipment until eventually the grid shut down. Why did this happen?

The authors provide us with a plethora of villains. Did a transmission line snap? Did air conditioners in the Midwest demand too much electricity? Maybe sabotage?

The authors place the blame largely on a company called First Energy in Cleveland, whose power plant didn't respond quickly enough to power fluctuations in Ohio. They also condemn the network of power plants that supply electricity to the grid. They operate too independently and communicate too poorly to control the gird.

The key question for most readers, though, when they finish this compelling report is this: Will there be a sequel? The authors say "yes" because a regional power plant may again fail, thus bringing down the power grid. Let's hope that it rises again like the Phoenix.

Text Messages

I'M A DEDICATED VIEWER OF the *American Idol* television show. In a case you've been in cave the last five months let me explain briefly: On the show contestants from around the country compete in a singing competition to get a multi-million dollar recording contract. Every week the audience votes, and the lowest vote getter leaves, until, by the final show, only one contestant remains standing. The by-product of all this, of course, is the highest-rated TV show of the season.

In that show I've seen the future—not of Pop Music, but the future of American cell phones. You see, viewers vote by calling a toll-free number, or by sending a "text" message with a cell phone. In that text messaging lies the future.

U.S. companies are trying to get Americans into the habit of using text messaging. AT&T tied in heavily with American Idol to get our fingers working: Viewers sent a record 13.5 million text messages during its five-month run. A drop in the bucket, though, compared to the 30 billion a month sent by Europeans.

U.S. wireless companies hope to duplicate the success of their European counterparts—although one European telecom executive theorized that text messaging will never become popular in the U.S.. He thought that Americans have eaten "too many hamburgers" and thus have "fingers too fat" to type the text.

Still, text messaging represents a cash cow to cell phone companies. For them that age-old aphorism "talk is cheap" needs to be amended with "text is cheaper." One trade journal reports that text messages have a profit margin of 95%. In fact, the UK cell phone giant Vodaphone gets 20% of its revenues from text messaging.

The cheapness of text messages lies in what's called, in engineering parlance, its asynchronous nature. That means that the sender and receiver don't need to communicate simultaneously, unlike voice calls which demand instantaneous transmission. The telecoms present text messaging as the latest high tech add-on, although it only requires old technology: It runs just fine on the current cell phone infrastructure. They nestle the text messages in small bits and pieces right between the technically more demanding voice calls.

The cellular companies would very much like us to be at the billion message a month level because they want us in the habit of using our cell phones for more than talking. They're building higher-capacity networks that allow cell phones to work as live picture phones, able to show video and to download music.

What's next, though, after this explosion of service? What follows every advance in communications: SPAM—annoying ads will now go mobile and be with us every step of the day!

Los Angeles

I'VE JUST RETURNED FROM one of the greatest engineering achievements of all time: Los Angeles. Yes, that's right, the city of Los Angeles.

To see the magnitude of this achievement I took a trip to the Getty Museum. It sits atop a mountain overlooking the city. From there you can see that Los Angeles sits in a semiarid coastal plain—the Pacific Ocean borders its west side, but desert boxes in the other three sides. Not surprisingly LA gets only fifteen inches of rain a year—yet the region supports over nine million people. Los Angeles grew from a city described in 1860 as a "vile little dump" to the second most populous city in the United States. The key to its growth: Water. In supplying that water lies great engineering. In fact, an achievement greater than the famed aqueducts of Rome.

To see the source of LA's water, travel north on Highway 395, which climbs up the eastern slope of the Sierras. At about 4,000 feet above sea level, just past the small town of Independence, you'll find two 20-foot long concrete blocks. They divert the Owens River toward the south—its the gateway for the Los Angeles aqueduct, a 235 mile canal that pipes water to LA. This mighty aqueduct was the brainchild of engineer William Mulholland. Mulholland emigrated from Ireland to the United States in 1870 at age 15. While living in Pittsburgh, he read a History of California that excited him so much that he had to visit the place. He traveled by sea to Panama, then walked 47 miles across the isthmus to catch another ship—he didn't want to pay the twenty-five dollars in gold required for a train ticket.

When he arrived in Los Angeles in 1877 he worked first as a ditch digger. The engineering of the ditches fascinated him. So after work he went to the library and studied books on mathematics, hydraulics, and geology. Self-trained as an engineer, he rose to head the Los Angeles Water Company.

His greatest feat was the Los Angeles Aqueduct which diverted the Owens River to they city. He supervised 5,000 workers as they dynamited and dug the 235 miles of canals and tunnels, laying it through sections of desert, and at times going through solid Sierra rock.

The Aqueduct opened November 5, 1913 in a spectacular ceremony at the north end of the San Fernando Valley. William Mulholland with his characteristic brevity said of the newly flowing water: "There it is: take it."

Angeleos still do: It carries 315 million gallons of water a day to Los Angeles. Yet, it is a most unheralded achievement: The only marker you'll find is etched into concrete walls hundreds of miles north of LA., but of course Mulholland's real monument is the city of Los Angeles.

Plasmas

A T LUNCH RECENTLY AN ENGINEERING colleague of mine claimed that the material world exists because of something unknown to most people. When I asked what that would be, he gave me a one word answer: Plasma. "Plasma," I said, raising my eyebrows, "what exactly is that?"

I learned that plasma is a fourth state of matter—beyond the three types we've learned about in school—the solid, liquid, and gases that we encounter in our daily lives. Plasma lurks around quietly in the background with little notice, yet it makes possible our interaction with these common forms of matter.

A plasma is an electrically charged gas. Usually a gas is electrically neutral, that is, the positive and negative charges reside together, but in a plasma they are ripped apart. So, in a sense a plasma is a charged cloud. In fact, a lightning cloud is a kind of plasma. Now how could such a thing affect our lives? Well, through the material things around us.

The fluorescent lights above our heads is plasma in action. When you switch on the electricity it separates the charged particles of the enclosed gas, thus setting up a controlled lightning display.

The tiny silicon microchip that runs every computer depends on plasmas. The circuit drawn on that chip is created by plasma etching: The highly charged gas burns very precise lines in the chip creating the circuit.

And plasmas appear in our lives in even more mundane ways. As my colleague and I finished lunch he shook a potato chip bag at me and said "plasmas even make this possible." Indeed, its very difficult to get ink to stick to a mylar chip bag—but blast it with a plasma and ink readily adheres to the bag.

Then he pointed out the window. "Ninety-nine percent of the universe is composed of plasmas," he said. The Northern lights, for example, are the emission of a glowing plasma.

So, the irrelevant detritus of our world—potato chip bags and the like—to the aurora borealis all depend on plasma. Now that is truly a cosmic oneness.

SpaceShipOne

ON MONDAY JUNE 21ST THE PRIVATELY owned SpaceShipOne flew to 62 miles above the Earth, creating the first non-governmental astronaut. Was this new astronaut's goal, in the words of Captain Kirk, "to explore strange new worlds. To seek out new life and new civilizations?" No, not at all. This newest astronaut and his private space ship pave the way for *you* to go into space!

Imagine the ride. A jet called the *White Knight* carries your space ship to an altitude of 50,000 feet—about double that of a commercial jet. Then the White Knight releases your ship, as it glides away the pilot fires the ship's rocket engine for eighty seconds and SpaceShipOne rises vertically to 62 miles above the Earth. You'll experience about three minutes of weightlessness and be able to see the black, starry sky unmediated by the Earth's atmosphere, whose thin blue line you can see below. The flight ends with a twenty-five minute glide through the Earth's atmosphere until you land in the Mojave Desert.

How much would you pay for such a ride? The entrepreneurs behind this and other private space craft are betting you'll cough up 100,000 dollars. They chat on about the need for private exploration of space—nonsense about "new energy sources on the moon" and blather about the precious metals stored in asteroids—but their real focus is a multi-billion dollar space tourism industry.

This might seem odd, even less than noble, yet it's the story of most forms of transportation. Showmen, for example, dominated ballooning in the 19TH century, selling rides at carnivals. The bicycle began as a fashionable novelty, rather than a practical means of transport. And the jets that we ride routinely have their roots in the "barnstorming" air shows where a visitor could buy a ride. Even the legendary pilot Charles Lindbergh began his aviation career by thrilling crowds as a wing walker.

So, the motto for the new private space flights won't be the Star Trek mantra "to boldly go where no man has gone before," but instead "to boldly go where all of my richer friends have gone before me."

Landlines & the Five Nines

WHEN MY FRIENDS TELL ME THEY'RE getting rid of their home phones—called a land line—and moving only to their mobile phones, I look them in the eye and say sternly "Remember the five nines!"

I'm not referring to any kind of numerology, but am alluding instead to the holy grail of reliability used by all engineers: The five nines refer to the numbers in ninety-nine point nine nine nine percent. The land line telephone network meets this standard: It operates 99.999% of the time. This translates into only five minutes of downtime in a year's operation. Contrast this to the 98% reliability of the cell phone network: That 98% sounds like a lot, but it means that for 10,500 minutes of every year a cell phone won't connect!

I know that it seems a brand, spanking new technology like cell phones should beat the heck out of a 19TH century technology like land line phones. Yet those 19TH century roots are the source of reliability.

For example, ever notice that in a blackout your phone stays on? Since the telephone system rose before a comprehensive power grid covered the country, the telephone company developed their own batteries that power your phone from the central switching station. In contrast cell phone towers depend on the power grid: No power to the tower, no service to the cell phone.

Through 100 years of slow and steady growth the telephone system has developed extremely reliable switches, even entering the digital age. Yet, again, a major source of reliability is a 19TH century one: Human intervention. Yes, people still monitor the network's operation. In fact, many telephone outages are ended when a human sees something wrong and simply throws a switch.

Yet landlines are disappearing. In the coming year some 19 million people will drop their home phone lines. A new Federal Law allows cell customers to keep their number when they change wireless providers. This has enticed millions to "cut the cord" and rely only on cell phones. Will the cell phone network become as reliable as the land line network? Perhaps not.

You see right now the FCC and the Homeland Security Administration disagree about whether to make cell network outages public. The FCC says "yes," crediting the public reports of the land line companies with dramatically improving network quality. Homeland Security says "no", outage reports would be a blueprint for terrorists who want to disrupt the U.S. They would provide a road map for targeting network stress points and vulnerabilities.

Perhaps. But if we all just keep our 19TH century phones, we could defend our 21ST century communications from terrorist attacks!

Tour de France

LANCE ARMSTRONG HAS AGAIN dominated the Tour de France. No doubt he's a great athlete, but he commands the race partly because he uses every high tech improvement allowed by the rules.

I know of no more high tech sport than cycling. Riders win the Tour de France by the tiniest of margins. In twenty-one days of riding covering 2,110 miles, the victor has won by as few as eight seconds!

To shave seconds the riders turn to high tech. Even though the International Cycling Union rules spells out the type of equipment allowed. The race "asserts the primacy of man over machine"—there's still room for a great deal of fancy engineering.

Engineers help cyclists battle drag—the wind resistance that impedes forward motion. No longer does the cyclist put the pedal to the metal—there isn't much metal left. Bike frames are made of aerospace-grade carbon fiber. The engineers have carefully arranged the fibers so the bike is rigid laterally for hard peddling, yet flexible vertically for comfort. The carbon fiber makes the bike lightweight: The rules allow a minimum of 6.8 kilograms, about 15 pounds—and no cyclist worth his or her salt uses anything heavier. And where the bike designers still use metal—the gears and the bike chain—they've hollowed out the aluminum crank arms and replaced much of the gears with ultra lightweight titanium. The high tech of the frame doesn't end with the material.

The shape of the frame reflects extensive wind tunnel tests as do the wheels. Engineers shape the wheels so that air flows over the tire and wheel smoothly without breaking into a turbulent mass that causes drag. They added dimples to the rim like on a golf ball: These keep the air flow attached longer reducing the drag-causing eddies that form. Even the clothing has a chest vent that sucks in air when upright, passing it around the body and closes when the rider is hunched. And the shoes that move the pedals went through 100 prototypes before the designers settled on synthetic leather that won't stretch or shrink in rain or heat.

And just riding the bike makes use of high tech gizmos. Lance Armstrong uses a computer-based power meter to measure every aspect of his cycling—track speed, heart rate, incline, cadence, altitude gain and power expended—so that next time his performance can be optimized even more.

What's left, you ask, for the cyclist to do? Well, pedal ... so far the rules outlaw motors.

Nanotechnology

WITH NANOTECHNOLOGY engineers manipulate atomic sized particles to create tiny machines. They'll be able to create, for example, toothpaste filled with nano-particles that repair damaged teeth, or pills that are really tiny pacemakers. Although still a young technology, the National Science Foundation forecasts the U.S. Market will be one trillion dollars by 2016.

Yet this promise may never be fulfilled, but not for lack of technological know how or resources: The U.S. Government alone will pour 3.7 billion dollars into nanotech over the next four years. But to really thrive a technology needs more than a scientific side, it must fit into our world socially and legally. For nanotech storm clouds already loom on the horizon.

For example, Britain's Prince Charles suggested that nanotechnology could be a disaster like thalidomide—the drug that caused grotesque birth defects in the 1960s. His remarks signal to the nanotech community the work yet to be done in creating a public receptive to their technology.

I suggest they look carefully at two negative role models: Biotechnology and nuclear power. Nether industry conveyed to the public the benefit of their product, nor did they listen to public concerns. In the absence of intelligent dialog, heightened concerns grew over the risk, nearly crippling both industries. Better public engagement could have prevented this backlash. The public isn't going to accept any technology where there hasn't been detailed studies of risks and benefits.

Right now the lack of information about nanotechnology invites alarmist scenarios. The nanotech industry needs to educate the public about what exactly nanotechnology is, and it needs to listen carefully to public concerns.

And there are other ways nanotech needs to fit into our world before being fully accepted—consider legal and regulatory aspects. The EPA is deciding whether to regulate nano-materials under the Toxic Substances Control Act, or to classify them as the naturally occurring "ultra-fine" materials—the same as dust, forest fire smoke, volcanic ash, bacteria and viruses.

And Patent Examiners are grappling with nanotech. If you use nano methods to make a tiny motor is that legally any different than a full-sized motor? In the past a simple change in size hasn't been patentable absent some other utility or novelty that comes from miniaturization.

To researchers who enjoy conquering the technological problems of creating a nanotech world, these social, legal and regulatory concerns may seem like dull things. Yet, some fraction of the nearly four billion dollars being invested into nanotechnology needs to be used to answer these questions. If not then these tiny nano-sized machines will bite back big time.

RFID

WAL-MART KEEPS TRACK OF ITS inventory using something called an RFID chip. This small microchip allows products to be recorded automatically by a radio receiver. The RF in RFID stands for radio frequency. Many roads use the same technology to keep track of tolls. And oddly enough, the Mexican Government uses it to track its Attorney General and his staff—all 160 of them. Yes, that's right, they are putting these chips into people.

Doctors implant them just under the skin, usually in the arm. The simple procedure feels about like getting a shot. The doctors never remove it, nor are there batteries to replace: The chip lies dormant under the skin until read by the radio receiver, drawing energy from the radio waves. In the case of Mexico, the chips let the Attorney General's staff roam the secure areas of his office without flashing a badge—the implanted chip grants them access.

These devices aren't approved yet in U.S., and I hope they never will be. This implant seems new and revolutionary, but it's just one more step down a perilous path we've been taking since World War II.

We feared at the end of the war the world of George Orwell's *1984*. But it isn't Orwell's Big Brother Police Force and their in-your-face technology that menaces us. Since World War II we've moved step-by-step toward a system where a police state need no longer be brutal, or openly inquisitorial, or even omnipresent in public consciousness. Police have instead moved in the direction of anticipating and forestalling crime. So, the trend is toward tracking every citizen throughout his or her life—geographically, commercially, and biologically.

This began soon after World War II with records of fingerprints, extensive paper dossiers on citizens, and then computer punch cards to sort through files. It evolved into the electronic databases and biological profiling we have today. These new chips are just a way to quietly add a page to an electronic dossier.

Still, the potential for abuse is enormous. In the future, perhaps, when someone approaches a sales desk their credit info would be displayed automatically for the sales staff. Or, the state could track the public movements of everyone. As a result people would be less likely to do public activities, to engage, for example, in protests that offend powerful interests.

So, at first implanting a tiny radio-frequency chip may seem a painless way to keep order. But we should keep in mind the story of police work since World War II: The most insidious technique is the one which makes itself felt the least, and which represents the least burden, yet lets every citizen be thoroughly known to the state.

Olympics & Technology

I WATCHED THE OLYMPICS THIS YEAR, thrilled to see athletes break world records. Now that the medals have been awarded I celebrate the unsung hero when it comes to setting world records: Technology.

Take time keeping, for example. For the first thirty-six years of the modern Olympics the judges brought their own watches. This mishmash of precision made it murky whether a record was really broken. In 1932 the Olympic Committee outfitted all judges with identical precision stop watches, and that year they even used newsreel footage to settle a disputed call. By 1964 the Olympics moved to quartz watches, which could measure a 100TH of a second. From there the timing keeping and measurement has only got more high tech.

The Olympic committee now employs full time engineers to design and improve the timing devices. The uprights of the pole vault use light emitting diodes to determine the height an athlete rises. In the swimming pool the racers slap a special pad at the end of the pool shutting off their clock. It's specially designed so wave action doesn't affect it. And sprinters cut through a light beam at the finish line, which measures the time to one thousandth of a second.

This super accurate time keeping allows records extended as the measure of a performance becomes finer and finer. It works like Zeno's Paradox where the tortoise and the hare cut the distance they race in half every time and never, in Zeno's world, reach the finish line.

But the story of sports records and technology doesn't end there. The height of a world record pole vault has got higher and higher as the pole changed from bamboo at the turn of the century, in the 1950s, and then glass-fiber composites today. Since 1912 the world record has increased by 53%, compare that to the 18% increase for the long jump, which doesn't benefit that much from high tech improvements. And in 1996 a new kind of blade revolutionized speed skating—and even the ice is different: Sophisticated temperature and humidity controls keep all the ice in prime condition for maximum speed. This control of the environment extends to other sports: For example, swimming has changed because today engineers design pools to minimize wave action.

And lastly even training has changed. Today athletes in endurance events measure the amount of oxygen their bodies use, thus allowing them to train optimally.

Now, with all this talk of technology I don't mean to take away from the athletes' superb performances, but I would like a change in the Olympic motto. Since 1896 it's been Citius, Altius, Fortius, which means swifter, higher, stronger. I'd like it to become Citius, Fortius, Altius Techne, that is swifter, stronger, and higher tech.

Betamax vs VHS

ONE OF THE MOST COMMON QUESTIONS I get is why did Betamax fail when it was so much better than its competitors. Only if you're born before 1960 does the word "betamax" even ring a bell. It refers to one of the earliest VCRs.

Sony came up with the Betamax, and JVC invented the VHS VCR, which became the standard we use today. Many consumers thought the picture quality of the Betamax superior to that of the VHS machine, so the Betamax got dubbed the "better" machine. Yet to my engineers eye, the better engineered machine was the VHS. Here's the story.

In the 1960s JVC and Sony each tried to make home video recorders. When they failed both companies returned to their drawing boards. JVC reinvented their VCR by asking themselves: What went wrong with the first home recorder? To answer this they asked what assemblers, retailers, repairmen, customers—everyone—wanted in a VCR. They wrote a plan they called "The Matrix" that listed the needs of everyone involved with the making, sale or use of a VCR. This guided the design of every aspect. For example, from this "matrix" they determined that the VCR must have a minimum recording time of two hours because consumers wanted to watch feature length films.

From 1971 to 1974 the engineers worked quietly to meet the goals of their matrix, then the JVC Chairman asked to see their work. They explained to him the nuisances of their video recorder and then popped in a tape. After the demonstration the Chairman smiled, leaned over, pressed his cheek against the recorder and said "It's marvelous. You have made something very nice."

Next they had to go up against Sony's home video system. Sony invited the JVC engineers see their new Betamax VCR in hopes that they'd all agree on a standard. They'd schedule the Betamax to hit the domestic market six months ahead of JVC's machine

Worried that the Betamax would take over the market, the JVC engineers carefully studied the competitor's VCR. They noted right away how well their matrix had served them. The Betamax tape was only one hour long! And the Betamax weighed some 30% more and it was more difficult to manufacture—the cost of manufacturing scales directly with weight. Their VCR might not have the picture quality of a Betamax, but it did fit the bigger picture: They knew it could meet the needs of manufacturers and consumers.

Soon after its release VHS sales caught up to Sony's Betamax, eventually surpassing it until now VHS is the worldwide standard. All because JVC engineers sat and thought and thought before they built a thing. So while the Betamax had a sharper picture, that's only a small part of the story. The rest of the story includes everyone who uses a VCR and that's where VHS beat Betamax.

Public Key Encryption

As an eight year old I spent many enjoyable hours with a book on codes and ciphers, but in today's world cryptography is no longer child's play. Our networked, computer world depends on it. Everything from a credit card purchase on the web to international banking. This networked financial world runs only because of a revolutionary breakthrough in cryptography.

In the mid-1970s a handful of young computer wizards invented a new way to encode information. They were distrustful of the government since they'd lived through Vietnam and the Watergate scandal. So, they thought deeply about ways to preserve privacy in the emerging computer age. They solved an age-old problem with codes: How to tell the recipient of the coded message how to read it.

Julius Caesar, for example, sent messages to his generals encrypted by shifting the letters of the alphabet a fixed amount. In order for his generals to read the messages Caesar also had to send along the key by a separate courier, which could be intercepted by enemies. Think of how cumbersome this would be in our internet age.

Consider using your credit card to purchase something from a web site. If you encrypted your card number the way Caesar did, you'd need to also ship the key to the web merchant. It wouldn't be safe to send in the open via email, so you'd have to do something like to drive it there, or mail it, which might not be secure.

To overcome this these young mathematicians invented public-key encryption, which breaks apart the coding and decoding. Here's how it works: I list my public key anywhere—maybe on my web site. With this you can encrypt a message to me, but here's the special part: You cannot then decrypt the message. In fact, you don't need to give me any special instructions about how to read the message. I have instead what's called a "private key." I'm the only one that needs to have this. This truly revolutionary separation of encoding and decoding makes possible our web-based internet world. For example, your web browser uses a merchant's public key to encrypt credit card information, then the merchant uses their private key to read it. No mailing of secret stuff needed at all.

Now, of course, like all technological developments this has a down side. If you can send your credit card information so that prying eyes can't view it, then a terrorism group can plan with the same degree of privacy. This has put cryptography under attack from the U.S. government for years—including banning certain software and even prosecuting its developers. Cryptography seems so exotic, and so much the domain of spies and secret agents that we lose sight of how essential it is to our modern everyday world. So, like all measures against terrorism, it abridges our freedom whenever we restrict the use of encryption.

Apple's Killer App

IN SEPTEMBER CBS NEWS retracted a claim that they had documents describing George Bush's National Guard service. Their news anchor, Dan Rather, announced they were forgeries. I've waited for the political fervor to subside to share with you what these documents really reveal: It isn't anything about politics, or even journalistic ethics—the documents capture the fusion of three events that gave us the age of the personal computer. In fact, the clues to the story are captured in the spacing between the letters.

Unlike a typewriter, which gives each and every character the same amount of space, the spacing between letters in the forged documents differed—the letters "t" and "a", for example, were closer together than the letters "s" and "p." This is called "proportional spacing"—a concept that was central to the rise and ubiquity of the personal computer, especially the early 1980s Apple.

Sales were sluggish at first for the computer, until a "killer application" came along—that's the computer industries lingo for a piece of software so powerful and so essential that everyone needs to have it. That software turned the computer into a sophisticated printing press. Just as the Apple computer appeared, two researchers founded a company called Adobe and developed the laser printer. The key to this printer lay in a revolutionary computer programming language called "Postscript", which allowed the PC to control every aspect of a printed image. Central to their vision was developing typefaces—a typeface, after all is nothing more than a very carefully executed drawing.

Now, it may seem odd to focus on letter spacing, but that's been a holy grail of the printing industries since its beginning. Early type setters pictured themselves as weavers of words and their text setting machines like looms. They sought to weave the text as beautifully as possible: No careless spacing of letters, but to create instead a block of text that appealed to the eye -- in fact, some think the name for a written page came from *textus* which means cloth.

Now neither Apple nor Adobe would have amounted to much without a third event. A small start up company named Aldus created the program called *PageMaker*. This turned out to be the "killer application" that ignited the PC revolution: It gave birth to desktop publishing. Within a year this trifecta of laser printer, *PageMaker* and the personal computer revived the struggling Apple brand and turned Aldus and Adobe into rich companies.

Together they created a graphic revolution that rivals in impact Gutenberg's printing press. So, we may overlook, as did the forger of the National Guard memos, the matter of a fraction of a hair's worth of spacing between letters, but in doing so we skip over the spark that ignited the PC revolution.

Container Ships

YESTERDAY I WALKED THROUGH AIRPORT security in San Antonio and got thoroughly searched because I didn't want to take my shoes off. As the earnest security guard searched me, I wondered whether he realized there was a much larger security hole than the shoes of a middle-aged man in a south Texas airport. Specifically, I mean the global supply chain that feeds the American economy.

Today retailers like Wal-Mart no longer have real warehouses, and manufacturers like General Motors don't stockpile parts for their assembly lines. They use what's called "just in time" delivery of stock and parts. They rely on a global supply chain, which depends on the very prosaic, yet astonishing technology called a container.

These containers come in only two very precisely regulated sizes: A 20 foot length, and the more common "double size" which is 40 foot long—both are eight feet in height and width. It's hard to picture their size, but consider this: You could likely store all your household goods in a 20 foot container. Although they're large, picture these containers being handed off like footballs: Those coming from Asia, for example, stop in nearly twenty ports before arriving here. They move very quickly because that's how the shipper's make money.

One 40 foot container filled with shoes costs about $5,700 to travel from California to Belfast. That may seem like a lot, but a 40 foot container holds precisely 12,384 shoe boxes—enough for a new pair of shoes every day for 24 years. This, of course, is only one container on a ship that holds up to 3,000.

Some 16 million containers travel around the world carrying 90% of the world's traded cargo. About 9 million arrive in the U.S. via our ports, or over our borders by truck and train.

In that astonishingly large number lies the security problem: With containers coming from all corners of the world, a terrorist could hide something like a dirty nuclear bomb.

To search one of these containers takes three workers about five hours, which makes it impossible to control the millions that go through our ports. In a few ports we have x-ray equipment that scans a container in 15 seconds—but only 4% are searched this way. To install the technology in every port would cost a half a trillion dollars.

Our only countermeasures are to develop cheap methods to detect the contents, and also to require twenty-four hour notice of the container's load, and its route. This would allow the U.S. to flag certain ones for special scrutiny.

So, if we think that the 9-11 terrorist attacks which shut down our air travel system stopped the country, think what would happen if the global supply chain -- all those container ships -- stopped moving for even a day. In short order the Wal-Mart shelves would empty and the assembly lines would halt.

Voting machines & gambling

AT THE SAME TIME AS AMERICANS go to the polls, Australians head to the horse races. The first Tuesday of every November they watch the Melbourne Cup, betting 100 million dollars on its outcome. It's too cynical to think of our elections as a "crap shot" like a bet in a horse race, yet the two events have much in common. It was a voting machine that made both activities as honest and above board as possible. The story begins in Australia with an engineer named George Julius.

Julius inherited from his father, an Anglican Archbishop and a fiery, anti-gambling crusader, a bent for tinkering. He learned from a friend of voter fraud with paper ballots. So, Julius tinkered until he'd invented a voting machine that allowed no vote rigging. Proudly he submitted it to the Commonwealth Government. This tamper-proof machine, of course, scared politicians and so they rejected it.

Soon after this a friend took Julius to the race track, a place he'd never been before. He saw something called the *jam tin tote*. A metal can where bookies placed paper bets. At the end of the race an army of clerks had to tally all bets—at some tracks there could be a hundred thousand—and then distribute the winnings in proportion to the wagers. Julius realized his rejected voting machine could speed up this whole process. So by 1913 he'd changed his voting machine into a totalizator, nicknamed "the tote", which counted all the bets.

Most track owners thought this new machine would soon disappear because the first one seemed entirely too cumbersome: The machine occupied a whole room, filling it with a tangle of piano wires and pulleys to move its various wheels and gears. Yet by 1970 with few exceptions, every major racing center in the world used these Australian Totalizators. The last one being decommissioned in 1987 when a dog-racing track in North London replaced it with, of course, a computer.

Although computers are taking over the world, let's take a moment to celebrate George Julius's modified voting machine that helped make gambling popular around the world as a government-approved, legal pastime. And also keep in mind that although its mechanical relative, the voting machine, gets bad press nowadays, it did help clean up and standardize elections. If only, though, it could make voting as popular as Julius's totalizator made gambling.

Nylon

L AST NIGHT I PACKED FOR A week long trip, which reminds me to talk to you about my underwear.

It's made from the most marvelous and amazing synthetic fabric that dries in just a few hours after a wash. I have all sorts of clothes made from this material, which I use when traveling. This means I need to bring only a few day's worth of clothes because I can wash them quickly in the hotel. In fact, I once traveled to Paris for two weeks with only a small carry-on suitcase.

Because I'm an engineer I searched, of course, for the roots of this synthetic clothing revolution. The granddaddy of all synthetic fibers is nylon. The story begins in 1927 when DuPont tried to hire eminent chemists—including the University of Illinois's Roger Adams. He, like all the others, turned DuPont down, but he suggested they hire his former student Wallace Carothers.

Carothers's specialty was assembling molecules into long strings, called polymers. This interested DuPont for very practical reasons: A polymer is the basis for plastics. You can think about a molecule as a brick, and the polymer like a wall built from that brick. But for Carothers the practical stuff didn't matter: He just liked hooking together molecules.

At DuPont, Carothers used his special techniques to set the World's Record for the longest chain of molecules. Yet, Carothers wanted more: He wanted the longest chain allowed by nature. So, he asked his lab assistant to mix up a batch of chemicals, and to let it cook for twelve days.

Inside the brewing concoction the molecules linked and linked and linked until they'd formed a hard, opaque white substance. Carothers's assistant heated a glass rod and touched the white mess. As he tried to remove it, a gossamer thread stuck to the rod. Like taffy he pulled it out of the beaker. As he pulled the thread stretched an enormous amount and then snapped back. What a wonderful material—especially for making clothes! Carothers named the material "fiber 66", but, of course, that wasn't exciting enough for the marketing team.

They considered 400 names, including Duparooh—short for DuPont Pulls A Rabbit Out Of a Hat. They rejected that possibility and tinkered with the name "No Run"—like a run in a stocking—until it became Nylon. The announcement of the discovery of nylon created waves in the press. A *New York Times* headline blared: Chemists produce synthetic silk. Time magazine called it "as lustrous as real silk." DuPont built on this PR to mount an ingenious marketing campaign to push the nylon to replace silk in stockings. It made silk seem virtuous and nylon naughty. Very quickly, nylon mania gripped the nation—a predecessor to my own quick drying underwear mania.

Sears Tower

L AST MONTH AN ARCHITECTURAL organization officially declared a skyscraper in Taiwan as the world's tallest building. So, today I celebrate America's entry in the skyscraper contest: The Sears Tower, the tallest building in North America at over 1,450 feet.

This American masterpiece came from a very American tradition: An engineer named Fazlur Khan designed the tower. That doesn't sound very blue-blooded, but this Bangladesh-born engineer followed a very American custom: He immigrated here.

In 1955, newly graduated from the University of Illinois, he began working in Chicago ushering in a renaissance in skyscraper construction. This immigrant gave the most American of cities, Chicago, its distinctive landscape, including the John Hancock Building, and, eventually in the early 1970s the Sears Tower.

Sears, a retail colossus at that time, was outgrowing their headquarters and needed bigger ones. Unwilling to move to the suburbs and unable to buy blocks of downtown real estate they decided to reach for the sky. They turned, of course, to engineer Fazlur Khan to build them the world's tallest building.

Khan faced two significant problems in designing the Sears Tower: wind and money. As a building gets taller it becomes more flexible in the wind. To see this, think of a skyscraper as a diving board: If you stand at the end of the diving board: Your weight, like the wind on a building, causes it to flex. If the board is made longer, it will flex more under your weight, unless the board is made thicker. A skyscraper is just like this: The taller it is, the stronger it has to be to resist the wind. And here is where money enters: As a building gets taller, the amount of material needed increases more quickly than its height, and thus the costs escalate—at least by conventional methods. Fazlur

Khan found a clever way to cheaply build a skyscraper: make it from square tubes.

To illustrate his idea Khan would take nine cigarettes and squeeze them into a bundle, from the end it looked a bit like a honeycomb. Each tube was a building in itself, but bundling the nine together created great rigidity—and uses less material than conventional ways: Khan's Sears Tower uses about two-thirds the steel of the Empire State Building.

You can see these nine tubes in the Sears Tower. The first fifty floors are nine tubes laced together, following by floors made of seven tubes, then five tubes shaped like a cross, until the final ten floors of two tubes. This gives the Sears Tower its layered structure, which one critic called "a driftwood carving by some giant."

A carving that might well be from a dying breed. Skyscrapers first appeared when corporate giants built self-named monuments to house their workers—New York's Chrysler Building, Chicago's Sears Tower, San Francisco's TransAmerica Pyramid—but will they thrive in this new digital economy? Maybe not. Compare these corporate skyscrapers to the headquarters of today's financial colossus: In Redmond Washington, Microsoft's headquarters is only sixty-five feet tall.

Packaging

IN THIS GIFT-BUYING SEASON, you can study a sub-discipline of engineering that fascinates me: Packaging.

A good package blends technical know-how and psychology. The engineer must understand the chemistry between, say, orange juice and its coated cardboard package, but must also grasp the emotional chemistry between the package and the consumer. In Japan, for example, fish cakes come wrapped in paper that looks like traditional handmade leaves, yet it sports a bar code.

The grand daddy of all packagers is Gale Borden. Milk springs to mind when I mention Borden, which just shows how successful he was. He became interested in packaging food to make it portable after hearing in 1846 of the Donner Party.

While traveling west they became trapped by snow. Only 47 of the original 87 survived, and they did so by eating the others. This gruesome tragedy stirred Borden to action.

He first tried to condense the essence of meat into a biscuit. It took six years and sixty thousand dollars to develop, but failed because it was unsightly and unpalatable. But the idea of condensing fascinated Borden.

He advised his pastor to "Condense your sermons;" he told lovers to no longer write poetry, but to "Condense all they have to say into a kiss;" and he suggested you spend as little time as possible at a meal, he was always done in fifteen minutes. Borden's next attempt at food packaging came while on an ocean voyage to London.

The ship's seasick cows couldn't be milked, so the ship filed with the cries of hungry infants, which spurred him to invent condensed milk. Although experts told him it won't work, Borden put some in a vacuum pan and boiled it. It failed, just as the experts predicted, because it stuck to the pan. Being less knowledgeable than the experts, Borden simply greased the pan and it worked fine to make condensed milk. He canned the product, selling it with promise of all good packaging: a sanitary and safe product.

Borden's condensed milk took off with the Civil War because soldiers needed portable rations. In fact, wars are often the stimulus for a quantum leap in packaging.

From World War I emerged cellophane—it was used for gas mask lens—after the War packagers used cellophane to wrap and display goods. Then World War II sparked our current state of packaging by making America excel in plastics: Saran Wrap, for example, started in the War as a film to protect aircraft engines from water. And eventually World War II soldiers had plastic canteens and containers to protect ammo, and they carried rations in plastic or foil pouches.

Plastics were not the only innovation from World War II, it also introduced, of course, the nuclear age. I hope, for humanities sake, that this new type of war never makes for a packaging revolution.

Wind-Up Radio

MY WIFE GAVE ME AN UNUSUAL GIFT: A radio that runs off a spring. Its huge crank looks and sounds like a comedian's prop. Turn it fifty times, which takes about thirty seconds, and the radio runs for an hour. Why make such a thing in an age of batteries and electrical outlets?

The idea for this wind-up radio came to British inventor Trevor Baylis while watching a television program on the spread of AIDS in Africa. He learned the disease spread fast because of difficulties in communicating with remote villages. They lacked electricity to run radios—and also lacked batteries because they cost a month's wages.

So, this life-long tinkerer rushed to his workshop and designed a wind-up radio. He even built a working model using the spring from an automobile seat belt retractor. In a sense, the easy work was done, and the hard part was to come: To be successful an invention of this type had to be mass produced.

Baylis had to attract a manufacturer—and from them the tens or even thousands of dollars needed to pursue patent claims worldwide. After filing for a patent in the United Kingdom an inventor has only 12 months to file international patents to protect his idea—a very expensive process.

Baylis submitted his idea to manufacturers everywhere—and got rejected everywhere. Even provoking one expert to tell him that the spring, which powered the radio, wouldn't work: It would weigh 100 pounds and run for only ten minutes. This expert suggested Baylis run the radio with human heat from "under the armpit." Instead Baylis took a gamble: He went public with his idea, risking that someone might steal it.

He pitched his idea on BBC TV. Within four days he struck a deal with a South African entrepreneur—no British or American manufacturer expressed interest. Today 300,000 of these radios bring news and vital information about HIV, and other health matters to an estimated six million people in Africa.

Today Freeplay Energy, the company making the radios, is expanding their "low-tech" mission by moving into medical equipment. Since the infant mortality rate in Africa is often 30 times that of the Western world, they are focusing on equipment vital for neonatal care. For example, a fetal heart monitor that uses ultrasound to monitor how babies are doing in the womb.

And, it doesn't use electricity or batteries, but relies on a hand crank and a solar panel to generate energy—just like the radio my wife gave me as a gift. With one critical difference: In America wind-up radios are novelties, but in Africa, they can mean the difference between life and death.

Blue LED

L AST WEEK A JAPANESE COURT AWARDED the inventor of the Blue LED or light emitting diode—you know those tiny lights like on the end of a key chain—over eight million dollars for his invention, a piddling amount for an invention worth about 600 million dollars. Why such a fuss over a blue light?

These LEDs may even end the 100 year reign of Edison's incandescent bulb. Although a wonder of the 19TH century it's simply too inefficient for our age, giving off much of its energy as heat. Not true of the tiny LED that runs cold, plus, they will last nearly 100 times longer than a regular bulb.

All this was only promise, though, until a self-described "country boy", who worked at an obscure chemical company, made the key breakthrough. Not a country boy from Texas or Louisiana, but from Tokushima, Japan.

Shuji Nakamura knew that for years large electronics firms failed to make blue LEDs, but this didn't deter him. He worked for ten years, seven days a week, twelve hours a day to perfect a blue LED. His bosses complained, asking him to drop the project noting that the big players in the field couldn't even make one. Undeterred, Nakamura succeeded.

As a result of his work Nichia Chemical sold 580 million dollars worth of these blue LEDs—they, in turn, gave Nakamura a mere $165 bonus. And then he did a very UN-Japanese thing: He sued his employer.

Japanese corporations rarely share profits or patent rights with their engineers and scientists. But last year a Japanese court ordered a food manufacturer to pay an inventor a million bucks, and it forced Hitachi to also shell out over a million to an engineer.

And on Monday the Tokyo High Court approved a settlement that paid Nakamura, the inventor the blue LED, a record 8.1 million dollars.

Don't read this yet as a great change: Mr. Nakamura wasn't satisfied with the amount, he wanted the 200 million awarded by a lower court, but when it was turned over on appeal his lawyer advised him to take the offer since the probability of winning the suit was "zero."

The biggest winner of all might be America. Mr. Nakamura now works at the University of California-Santa Barbara. And he advises Japanese scientists and engineers "to come to America, where their abilities are reflected in their income."

Ferris Wheels

I CELEBRATE TODAY THE greatest engineer born on Valentine's Day. He made his mark in 1893 at the World's Fair in Chicago.

As the fair prepared to open, its organizers searched for an engineering achievement to surpass the Eiffel Tower. Today we think of it as a slice of turn of the century Paris, but at the time the tower was the latest in modern engineering. It stated boldly that the French were prepared to construct the bridges and buildings of the twentieth century.

So, the Chicago Fair's organizing committee wanted some distinctive feature. "Mere bigness," they said, would not be enough; instead, they searched for "something novel, original, daring and unique" to show the "prestige and standing" of American engineers. They already had a chocolate Venus de Milo and a 22,000 pound cheese in the Wisconsin Pavilion, but they wanted more.

A 31 year old engineer approached them with an idea. He'd been hired to inspect the steel in all of the Fair's buildings, and he wanted to build on that expertise to create a monument in steel. At first the committee turned down his idea as "outlandish" and "too fragile." Undeterred he wrote them a letter spelling out his plan: "I have on hand," he said, "a great project for the World's Fair in Chicago. I am going to build a vertically revolving wheel 250 feet in diameter." His name, of course, was George Washington Gale Ferris. His proposal became what we now call a "Ferris Wheel."

What a spectacular monument he built: Using 100,000 separate parts he created a wheel as high as the tallest skyscraper in Chicago— even higher than the Crown of the Statue of Liberty. The wheel's axle alone weighed 140,000 pounds, and its 36 cars, each the size of a railway car, carried 60 people. Yet it was the wheel's lightness that startled people: Its rim seemed to float in the air held up only by gossamer steel spokes, like a bicycle wheel.

Ferris created not only a monument to America's engineering prowess, but also a new aesthetic experience: As the wheel descended —it took about 20 minutes for a complete revolution—the whole fair ground near Lake Michigan slowly opened into view.

This ride succeed in bring notice and, more importantly, paying patrons to the World's Fair. Even today Ferris Wheels are used to celebrate big events.

For example, when the British needed to welcome the new century they created a 450 foot tall Ferris Wheel renamed, though, the Millennium Wheel. And engineers in Singapore are creating a 100 million dollar wheel that'll take passengers 560 feet off the ground, and give them a spectacular view of neighboring Malaysia and Indonesia. Thus meeting the standard for spectacle set by George Ferris with his Ferris Wheel in the 1893 Chicago World's Fair.

Bob Kearns

EARLIER THIS MONTH MANY newspapers printed a short obituary of an engineer named Bob Kearns. I want to take a minute to remember this engineer, and his struggle for justice.

Bob Kearns invented the intermittent wiper blade—the blades that flash occasionally across a windshield. This simple invention caused him much grief.

In a sense the impetus for the invention occurred on Bob Kearns' wedding night. He opened a bottle of champagne and the cork struck his left eye—forever impairing his vision. A few years later Kearns drove his car in the rain and noticed how the wiper's regular motion distracted him because of his poor vision. He realized that a wiper should be more like an eyelid, that is, it should blink occasionally. This thought spurred him to build an intermittent wiper blade, which he tried to sell to Detroit auto makers.

Now, wiper blades interested auto makers because wipers helped sell cars. In the late 1950s several cars began sporting two blades that swept in parallel across the windshield, replacing the single blade that created a huge Vee in the middle of the windshield. This new two blade system attracted buyers, and by the 1960s every car had them. The next step was intermittent operation.

Kearns' key invention was making a cheap timer for the wipers. The auto companies had developed a mechanical contraption to do this with some twenty-nine moving parts. Kearns design by contrast was elegant: He used an electric motor with a timer to control the wiper. The result: Four parts, only one of which even moved. But it seemed so obvious that auto makers thought Kearns' patents would be null and void.

So, they built cars with these wipers, but didn't pay Kearns. When he heard about this he bought a wiper and carefully took it apart: He saw all the essential parts of his 1964 patent. So he sued every major auto maker for one point six billion dollars—this was about 500 million for his lost profits and more than a billion in damages. The first auto maker he sued offered him thirty million dollars to settle out of court. But to Kearns accepting the settlement meant that it was OK to steal inventions.

His case went to trial after a twelve year delay—twelve years in which Kearns' single minded pursuit of justice lost him his wife and broke his health. After a three week trial the jury returned a verdict: The auto maker indeed infringed on Kearns patents, but they awarded him $5 million dollars—a far cry from the 1.6 billion he wanted.

So, the next time your wipers flash across your windshield pause for a moment and think of the memory of Bob Kearns and his struggle for justice.

Better Mousetraps

I OFTEN GET EMAILS AND PHONE CALLS asking "How do I market my invention." Most take to heart the aphorism "Build a better mousetrap and the world will beat a path to your door." When a reporter asked James Dyson, one of the few inventors who truly grew from a home workshop to a world-class industrialist, to comment on the truth of this advice, Dyson responded vehemently "no, no, no!"

The story of Dyson's success serves as a cautionary tale for all would be inventors. He invented a new type of vacuum cleaner, which has become the best selling brand in the United States and the United Kingdom.

It begins, as do many inventions, with frustration. As a six-year old Dyson had to help his mother with household chores. He noticed frequently as he vacuumed that he had to stop to clean out the sweeper. As with all vacuums the bag got clogged and the vacuum lost suction.

Dyson never forget this, especially some twenty years later when inspiration struck him while visiting a sawmill. He observed sawdust being sucked into a cone using a spinning column of air. He spent the next three years building 5,000 prototypes to invent a vacuum that worked on the same principle. A cyclone whirling at 900 miles per hours sucks up dirt and tosses it out the side—no filter, no bag, and best of all no lost suction.

Here indeed was something revolutionary. But that revolutionary aspect, Dyson pointed out, was exactly what creates the main problem for the lone inventor. For years he tried to get the big manufacturers —Hoover and Electrolux—to license his newly perfected vacuum. They declined: The revolution of the bagless vacuum was the last thing they needed because they sell $500 million dollars worth of bags for their cleaners.

So, for Dyson, like many lone inventors, he had to go it alone. He describes this as "fighting against this huge mountain that you've got to move." The mountain he refers to is not the public—they love new things, he notes—but what he calls the "man in the middle"—the distributors and shop keepers, who are faced with buying not one, but thousands of a new invention.

Small wonder he offers this advice to new inventors. The essential ingredient you are selling, he says, is "hope." You need to "inspire those in retail and manufacturing that your new, strange thing will really attract customers." That, in turn, is the advice I offer all who call me seeking advice on their revolutionary inventions.

Recycling

L AST MONTH I TRAVELED THROUGH the Pacific Northwest and California. At every stop I heard about Seattle's new tough recycling laws. Toss too much junk mail and cardboard in the trash instead of the recycle bin, and you'll be fined. Seattle makes mandatory the recycling of everything that can be recycled.

From an engineer's viewpoint, this doesn't make environmental sense. Now, before those nasty emails and letters start arriving, let me emphasize: I said it made no sense to recycle everything that can be recycled, I certainly didn't say all recycling made no sense.

Partly these laws arise because of a commonly held view that we toss, say, plastic soda bottles into the recycle bin and they go back to the manufacturer and they produce another bottle. In other words, a vision of a self-sustaining cycle of life. But the economy of recycling materials is more complex then this.

For example, only a bit of a recycled plastic bottle ends up in a bottle again. Instead it appears in carpets: Half of all polyester carpets come from plastic soda bottles. And there are other issues that determine the use of recycled materials. The cost of picking them up —it typically costs double that of regular trash. Greater costs means additional trucks, fuel and sorting facilities -- all of which take an environmental toll.

This critique, though, has the same flaw as the Seattle recycling laws: Its what an engineer calls an "end of the pipe strategy." In other words, only years after making something do we consider how to dispose of it. Instead we need to focus on "green design"—on designing consumer products from the get go to be reused or disposed. Here's an example. The two-liter soda bottle has a plastic label, which is different than the body of the bottle. So, to recycle this the two different types of plastic must be separated. In the past recyclers separated them with a simple flotation technique: The plastic in the bottle sinks, the labels floated. Recently the manufacturers put a wrench in the works: They changed the type of plastic used in the label to one that also sinks.

So, here are some simple rules for manufactures to incorporate green design, that is recycling their products: Make the components easily separate so no dissimilar materials remain together; make sure common tools like hammers, screw drivers, and pliers will suffice to get it apart; design it so one person can disassemble it in 30 seconds or less; and lastly, be sure each material type can be identified through markings.

So, to be environmentally friendly we want products designed so that the high value materials can be removed—reserving recycling for metals, using the plastic and wood products for energy-producing incineration, and then—and I know that this will bring in the letters —landfilling the rest.

Josiah Willard Gibbs

T HIS MONTH THE U.S. POSTAL SERVICE debuts a new commemorative stamp set to celebrate four American scientists. Over the next four weeks I'll share with you the achievement of each scientist. I begin with the first stamp: Josiah Willard Gibbs.

Although Josiah Willard Gibbs lived and worked mostly in the 19TH century—he made it into the 20TH only by a scant three years—he remains near and dear to my heart because I've taught and lived with his work for years.

Gibbs truly revolutionized the world, although the facts of his life suggest nothing interesting: Born in New Haven Connecticut—and died in the same place—in fact, in the family home where he grew up. He never married, and lived his whole life with his sister. He held only one job: Professor at Yale University—and even that he did for free for the first nine years. Only when another university offered him a salary of $3,000 did Yale starting paying him $2,000.

Gibbs studied thermodynamics—the study of how energy moves around. While at Yale he wrote an earth shattering paper with the cumbersome title of "On the Equilibrium of Heterogeneous Substances." As a young teacher of thermodynamics I decided to read it. I found it incomprehensible. In fact, in Gibb's own lifetime the paper was so confusing and so terse that it only had impact after a fellow scientist translated it into German and added commentary.

In that paper Gibbs taught the world how to think about chemical reactions. We're all familiar with them: They appear most vividly in the explosion of a firework. Before Gibbs chemists had to use trial and error to see if chemicals would react to form something new.

Here's what Gibbs did: Think for a moment of a car poised on top of a hill. Release the brake and the car naturally goes downhill until it rests in a valley; conversely a car at the bottom of the valley would never spontaneously go uphill, unless someone stepped on the gas.

Gibbs showed that chemicals behave in the same way: Either they'll react—that is go down hill together naturally—or, figuratively speaking, they're stuck at the bottom of a valley and won't react at all. Gibbs great genius showed chemists exactly how to calculate which reactions are poised on a hilltop, and which are stuck forever in a valley.

Does Gibbs deserve to be celebrated on a U.S. postage stamp? Yes. Thanks to Gibbs chemistry serves as the central science of our modern world, as the basis of nearly every manufactured object. The plastic case of a computer, for example, comes from mixing ingredients until they react to form a solid sheet. The same, of course, is true of everything else in your house: soap, glass, butter, and cosmetics, to name just a few. Rarely has such genius been observed in the scientific world as Josiah Willard Gibbs, and even rarer does it have this impact.

Barbara McClintock

THIS MONTH THE U.S. POSTAL SERVICE debuts a new commemorative stamp set celebrating four American scientists. Today I share the second in that series: Barbara McClintock.

No one knew more about a cob of corn than Barbara McClintock. Each spring this scientist rose very early in the morning to plant corn on Long Island Sound, carefully fertilizing each stalk throughout the summer, then harvesting them at the end of the season. She spent the long, quiet winter months analyzing her harvest. Unlike most scientists she worked completely alone, so much so that if a visitor showed up in the afternoon she often had to speak softly, saying she hadn't yet used her vocal cords that day. She studied the color of the corn kernels, which varied from dark to light. In the 1940s she noticed distinct and regular patterns in the colors of the kernels. McClintock knew that these patterns reflected the genetic make-up of the corn. A set of genes controls the appearance of every plant and animal; this is, of course, the genetic code that we use in DNA testing to determine paternity or to solve crimes. So, McClintock realized that the rapid change in the corn's appearance meant something revolutionary: No longer were genes the fixed, stable things always thought by scientists, instead they could spontaneously change.

She spent the next three years checking and double-checking her results before she revealed to her colleagues the existence of these "jumping genes."

What does McClintock's work mean for us today?

Her work led to greater understanding of human diseases. For example, how jumping genes can pass on resistance to antibiotics, or how they let African sleeping sickness evade the defenses of the human immune system.

Her work was so far ahead of its time that only 40 years after she did her ground breaking research did she receive a Nobel Prize. At age 81 all she had to say was "Oh Dear"—and then she walked out in to the brisk air of Long Island Sound and spent all morning picking walnuts. She returned, dressed in her dungarees and carrying tongs for grappling with the walnuts, to address the press. "It might seem unfair," she said, "to reward a person for having so much pleasure over the years, asking the maize plant [as she called corn] to solve specific problems and then watching its responses."

The world, though, is richer today because Barbara McClintock hear what the corn said back.

John von Neumann

THIS MONTH THE U.S. POSTAL SERVICE debuts a new commemorative stamp set celebrating four American scientists. Today I share the story behind the third stamp of the series, which features John von Neumann.

John von Neumann showed great mathematical skill at an early age: By six he divided eight-digit numbers in his head, by eight he did calculus and also demonstrated a photographic memory by reading a page of the phone book and than reciting it with his eyes closed—later in life he used that ability to memorize the world's largest collection of off-color limericks.

Although born and raised in Budapest, his great skill attracted the attention of the Institute for Advanced Study in Princeton, New Jersey. In 1933 they made him the youngest member of their faculty. His ferocious intelligence so awed everyone that they joked that von Neumann was indeed a demigod, but that he had made a detailed study of human beings and could imitate them perfectly.

Like many scientists of his age, he got involved in the Manhattan Project, which made the first atomic bomb. He solved a key problem with the early bomb designs. Nuclear weapons depend on conventional explosives to squeeze the fissionable material together so it reaches the critical mass needed for an atomic explosion. Von Neumann figured out how exactly to place these charges so they would initiate the nuclear reaction.

He couldn't, of course, actually detonate a bomb, instead he had to calculate what would happen. To do this he used a series of IBM calculating machines. Soldiers operating them formed a very slow human computer: As each finished his work, he passed the result on to the next person to use in their calculation. This work turned von

Neumann on to the idea of using computers for scientific research.

He next worked on the first real computer, ENIAC. He realized a limitation to its design: ENIAC had to be, essentially, rebuilt for every different type of computation. Workers actually had to unplug cables and reassign them to different jacks. This task often took up to 2 days. So, in 1945 he applied his powerful analytical abilities to solving this problem.

In what is perhaps the most famous paper in the history of computer science he spelled out his vision of a computer. He boldly drew comparisons between electronic circuits and the brain's neurons, emphasizing that just as the brain relies on memory, so the computer would depend on its programs. Computers, he thought, should not be specialized machines, but highly flexible, general-purpose devices, where a program, or what we now call software, controls the computer's actions.

So, while it's fitting that the U.S. Postal Service has dedicated a stamp to John von Neumann, his real memorial lies in the billions of computers sitting on desktops each doing its own thing because each runs it own special type of software.

Richard Feynman

THIS MONTH THE U.S. POSTAL SERVICE debuts a new commemorative stamp set to celebrate four American scientists. Today the final stamp in that series: physicist Richard Feynman.

Most Americans remember Richard Feynman either from his star turn on the panel investigating the 1986 Space Shuttle accident, or, from his surprise best seller *Surely You're Joking Mr. Feynman.*

Now from that book you'd think he was a curious character. That he only taught himself how to fix radios, pick locks, draw nudes, speak Portuguese, play the bongos and decipher Mayan hieroglyphics. All true, but they understate his achievement.

In 1965 Feynman shared a Nobel Prize for his work on something called Quantum Electrodynamics, more often known by its initials Q.E.D. It revealed the inner workings of electrons.

Why should we care about this? Feynman himself once said he didn't think it would ever amount to anything useful. His life and work ask the interesting question: Why should we care about basic research into how the universe works? And why, as a society, should we fund it?

Often we'll hear that basic research leads to technological innovation—to products on our desktops and in our homes. This is only a partial justification because sometimes it does lead to this, but sometimes it doesn't. The dance of science and technology is never one-way: The lead switches often.

Here's a second answer: Pure science is a great aesthetic and spiritual adventure for humankind, worth doing for its own sake like any art form. So, like symphony orchestras, we should fund this great romantic adventure—what more should a civilized society do than broaden the realm of human knowledge? In a society where we currently have great divides on cultural issues, its a bit hard to justify science as a cultural imperative.

To me the best answer lies in this: Basic research brings sharp minds to universities—minds which can train the next generation of engineers and applied scientists, and minds which have the time to think of something revolutionary.

So, basic research itself may not be productive in the utilitarian sense, but has value because it concentrates the best minds to create a rich, intellectual environment where bold ideas hold sway—and where the payoff for society may be very big.

Richard Feynman's career demonstrates this well. While teaching at CalTech he received acclaim for his work in physics, which revealed a fundamental understanding of the universe. That's all well and good, but he might well have thought up something that will impact lives every day. In 1959 he gave a talk entitled "There's a lot of room at the bottom"—in it he laid out a vision of the nanotechnology revolution that's occurring today. That speech has inspired a generation of researchers who may truly change our world.

Moore's Law

L AST WEEK THE CHIP MAKER Intel paid $10,000 for an original copy of a 40-year-old electronics magazine. It contained an article of faith for the company: Founder Gordon Moore's famous "Law" about computing power.

His wonderfully titled article "Cramming more components onto integrated circuits" predicted that about every year the number would double.

It's held up pretty well with computing power doubling every 18 months: In 1964 a computer chip had about 50 transistors, now they cram half a billion.

For the consumer this means that every year and a half a home computer becomes twice as powerful for the same amount of money. Also, Moore's Law applies to other computer-based devices: Hard drives and cell phones for example.

Chip makers remain sure they can do this yearly "doubling" for about five more years, but after that it's pretty dicey. No technology clearly stands out for continuing this explosive growth of computing power—cheap growth, that is. This means consumers will feel it in the pocket book if Moore's Law stalls.

They'll be stuck with the personal computer of 2010. And if they move, for example, into digital photography and advanced imaging, they'll find no cheap solutions for storing large numbers of photos beyond that age's capability.

And for the investor the center of gravity of the electronics industry will shift—in both hardware and software.

For hardware, advances in computing will come from outside the single-chip computer—the Intels of the world. The companies that thrive will be those that develop more intelligent peripherals, off-loading work from the computer's processor.

And Moore's Law gave us bad software: Instead of writing efficient code, programmers just waited until the computers got faster. I'm reminded of Blaise Pascal's comment three centuries ago. Apologizing for a verbose letters he wrote "I have only made this [letter] longer because I have not had the time to make it shorter."

Letters like software, take time to shorten and condense. So software giants like Microsoft find it better to write "just good enough" programs, that rely on advances by Intel to make them run efficiently. The current giants might be displaced by nimble companies that take the time to write efficient software. And when the age of Moore's Law ends, we may well be telling our children unbelievable tales about how in our day computing got bigger and better every year.

Housework

W E, OF COURSE, CELEBRATE Mother's in May, but it remains a sad fact of motherhood that women, particularly mothers, do the majority of housework.

In fact, earlier this month the Spanish parliament passed a law requiring men to do more housework. But didn't technology make all of this easier, nearly erasing housework?

No one disputes that at the beginning of the 20TH century it became much less backbreaking. In a sense just as the nation became "industrialized" so did the home: Electrical appliances and gas heating helped a housewife cook and clean. But for whom did appliances make work easier? And for whom did they save time? Consider the simple example of a meal. Most families in the 18th century ate stew —just a big pot of meat and vegetables cooked in liquid for a long time. To make the stew took both sexes.

A man used handmade knives to butcher an animal, a woman carried water to the house in wood buckets, held together by leather likely made by her husband. She cooked the stew, made from vegetables from her garden, over a fire using wood chopped by her husband.

She thickened the stew with grain husked and threshed by her husband. Any scraps or garbage that were not used were moved outside—likely again by her husband.

Now let's fast forward to today.

We buy food from the grocery, toss it in a manufactured steel pan, flick on a burner and cook dinner—disposing of the scraps down a garbage disposal. Look what happened to housework: Technology has liberated men from their role—no need to butcher animals, or cut fire wood, or even toss out the food scraps. Examples like this abound.

Look at cleaning a rug.

In the past it might be a children's chore to take the rugs outside a few times a year and beat them. Today, of course, mothers drag a vacuum cleaner across the carpet—no need for the children to help. These example give us a way to look at any technological innovation: Keep the new technology in perspective by looking at the big picture. For housework it didn't create more time, technology's main effect was a gender shift of work toward women. And in addition, these newer technologies created a higher standard of cleanliness for a housewife to achieve.

So, if you'd like to give your Mother a real gift for the holiday you can start by doing all the housework.

Local Power

IN THIS ENERGY CONSCIOUS TIME of ours there is one single thing we could do to conserve billions of barrels of oil: Move our power plants closer to us!

Since the early 20TH century we've built them farther and farther away from cities and towns. We learned with the first ones in the late 19TH century that no one liked having a smoke-spewing plant next door.

So, by 1910 or so power companies built them far from city centers. At first moving them made for economies of scale because one plant could serve several cites. Yet we paid a penalty: A decrease in efficiency.

By efficiency I mean the fraction of energy that enters compared to the amount of electricity created. You see, a power plant burns oil to produce electricity. Each gallon of oil entering has a maximum amount of energy, but we don't convert all of that to useful electricity. Some it goes into running the equipment in the plant, some we lose as heat. Also transmitting the electricity long distances over power lines involves energy loss as heat, typically about 10 to 15 percent.

With all that in mind, here's an astonishing fact: Our peak energy efficiency occurred in 1910. Yes, that's right 1910. We converted about 65% of the input energy into usable electricity. My moving power plants further away we dropped by 1960 to an efficiency of about 30 percent. Partly this occurred because of lost energy in long transmission lines, but also because the heat from the plant could not be recycled and used in nearby buildings, instead they simply vented it.

With today's technology we can convert more than 50% of the energy from burning fuel into electricity, while at the same time giving off fewer pollutants. New technologies also create quieter plants, ones that would be good neighbors. In addition we could recycle waste heat to nearby office buildings and homes, and with nearby plants we would lose less energy to transmission.

All combined we could achieve between 65 and 95 plus percent efficiency. Creating an infrastructure of these smaller, de-localized plants would mean a savings, over three decades, of $5 trillion in capital investment, and would consume 122 billion fewer barrels of oil. All this means, of course, less carbon dioxide emission, less pollution, and cheaper energy.

Open Source

For years Microsoft's Internet Explorer has dominated the web browser market—by some estimates capturing almost 100 percent. But their share has dropped to 89% and continues to decline. An upstart browser called Firefox recently clocked its 50 millionth download. More is at stake, though, then surfing the web.

The Firefox browser represents a new way to write software: No giant corporation, just thousands of volunteers around the globe linked via the internet. They are part of an informal revolution called the Open Source movement.

By Open Source I mean that their computer code is available to anyone to modify as they wish. This is a stark contrast to Microsoft, for example, that keeps it Windows operating system under tight wraps: Only Microsoft employees know how it does its magic.

It would seem that this new model doesn't build strong software, but there are thousands of programmers who will work on the code. They are hyper-picky people who enjoy finding errors and fixing them. The result to the user—and the main reason I use Open Source software—is that it's incredibly robust—it rarely crashes. In fact, it's the backbone of my computer network.

Microsoft would love to stop this development, but it can't follow its usual strategy of buying up a competitor: There isn't any company to buy, because Open Source software is the product of individual programmers all over the world.

But what really terrifies Microsoft is the license used on this type of software. It says that anyone can use the code, but—and this is the twist that scares Microsoft—if you do you must release your entire source code, and allow anyone to use it and distribute it. No wonder a

Microsoft Vice-president likens this license to a virus.

So, will the Open Source movement be the David that eventually slays the Microsoft Goliath? It's unlikely it'll dislodge Windows any time soon from your home computer, but I'll tell you where to watch the battle: On the internet.

Since 1995 the number of computers that route web traffic and e-mail around the world has grown from twenty-five thousand to six billion. Microsoft would love to dominate this market—yet over 70% of these web servers run an Open Source Program called Apache, compared to only 20% that use Microsoft software. And the Open Source program is increasing its share every year.

My prediction here turned out to be true: A unix ancestor runs Apple Computers, the dominant Smartphone platform is open source, and indeed the internet is the battleground for information technology dominance—just look at the rise of search giant Google.

Voice Over IP & 911

MANY USERS HAVE MOVED THEIR long distance phone
service from the standard land line to something called VOIP
—or "voice over internet protocol", sometimes called an "internet
phone."

As the name implies it uses a computer and a broadband internet
connection like cable or DSL to make phone calls. Recently, this
technology made headlines, and not in a positive way. Turns out many
users could not call 911 in an emergency. So, the Federal
Communications Commission voted unanimously to give the VOIP
providers 120 days to tie into the 911 system.

You might think this would be just automatic. I mean, what's the
problem? After all its just a phone call. Well, the internet phone
companies can't just flip a switch to add 911 because it's really a tricky
technological problem.

When you make a 911 call, a computer gives the emergency operator
your location. This alone presents problems for an internet phone.
You see, engineers, in a wonderful phrase, call the internet
"geographically agnostic." That is, a computer address exists no where
in particular geographically. A land line phone, in contrast, resides at a
specific location. Even cell phones have more location information
than you'd think because their signals go to a stationary tower. The
whole point of the internet, of course, is to de-localize a person: You
can be anywhere in the world and still use the internet.

Beyond this problem of locating a 911 call, linking to the current
emergency system presents big hurdles. Some of it stems from the
problems of fitting a computer into the ancient 911 network. Unlike
the phone system, which consists of one standard type of device, a
voice over IP 911 caller might be on a DSL line, a cable modem, or a

wireless or satellite broadband connection. A nightmare when interfacing with the 911 system designed 40 years ago.

The internet phone companies also have to overcome a political hurdle. By law only "telecommunication services" can access the 911 system, not "information services" as the Feds classify voice over IP. This would mean partnering with an existing telecom provider, who is also a competitor.

In spite of all this, we should not give up on internet phones. In the long run they'll be cheaper than regular phones and will offer enhanced services in an emergency. For example, notifying neighbors when 911 has been dialed, or giving complete medical info to emergency personnel.

I hope the 120 days the FCC has given internet phones to provide 911 service won't put them on permanent hold. Clearly for now one should have a land line nearby for emergencies, but stay tuned because internet phones are the way of the future.

Color Film Chemistry

A N OBJECT FROM MY YOUTH has disappeared: Photographic paper for printing from negatives. As a child I loved seeing a photograph appear slowly in a tray of developer. No more: With the digital revolution in full swing Kodak will stop making the paper this year.

Photography illustrated for me that scientists engage in creative acts, not just learning a list of facts. Take, for example, another Kodak innovation: Color film. It was invented by two friends who loved first and foremost, not photography, or chemistry, but music.

In 1917 two teenagers, Leopold Godowsky and Leopold Mannes, saw a blurry color movie. Dissatisfied with the image they decided, with the hubris of youth, to make a better color film.

At that time color photographs and movies were made by using three pieces of film, one for each primary color. These separate pieces were combined in a projector to create a full color image, yet always blurry because the three pieces of film never aligned correctly.

Godowsky and Mannes realized that to get sharp color images they needed all three layers to be part of the same piece of film. In their lab they found it very tricky to create a thin layer for each color, then to separate these with an even thinner layer of clear gelatin.

Although dedicated to making color film, music often interrupted their nineteen year quest. Godowsky studied violin at the University of California, and Mannes studied piano at Harvard. Then Mannes won Pulitzer and Guggenheim scholarships to study music composition in Italy.

By 1930 they found their experiments so complex that they could no longer fund them from their musical performance fees. The Research Director for Kodak had learned of their work. He hired them to work in the Kodak laboratories. So, they spent days in the lab, then evenings performing at the nearby Eastman School of Music, although the music wasn't completely separate from their lab work.

Godowsky and Mannes would sing as they worked in their labs, not for fun, but as an essential part of developing their color film. In their darkened lab they were unable to see a watch, so they timed the reactions by singing passages from their favorite musical pieces, whose length they knew by heart.

By 1935 they perfected their color film. In an odd press conference, the inventors announced their discovery, showed sample photos, and then played a violin and piano sonata for the reporters. After this they developed no more film, but instead returned full time to musical careers.

No surprise or discontinuity, here, though, because creating color film or playing a sonata are both creative acts. So, to continue playing music for them, was one and the same, at a fundamental level, with developing another type of film.

Jack Kilby

L AST WEEK AN ENGINEER who brought us the information age passed away. Jack Kilby died at aged 81 after a life filled with inventions and honors, including in 2000 the Nobel Prize for Physics.

I only met Jack Kilby once, and very briefly at that, yet for years before that I'd told people he gave me the gift of dance.

Six years ago my wife and I joined a ballroom dancing class with the slogan "Go from two left feet to the dancing elite." As the class began I danced so badly that my wife looked at me and said "What is this? Am I dancing with Woody Allen?" Determined to join this elite I bought a metronome—and that's where Jack Kilby's invention of the microchip comes into play.

I looked for the metronomes of my childhood piano lessons—large pyramidal objects with swinging pendulums. The music store clerk handed me, instead, a device no bigger than a credit card and with no pendulums, just press a button and a chime beeps a waltz rhythm.

I tossed this tiny device into my breast pocket and practiced everywhere: I danced in the hallways at work, down the aisle of the library and even turned a head or two on the street. And thus I joined briefly that elite group of ballroom dancing people.

Moving from that large mechanical metronome to a tiny credit-card sized device really captures Jack Kilby's impact. His microchip removed mechanical devices from our world, reducing the number of moving parts and dramatically increasing reliability.

For example, watches no longer depend on clockwork, but run off a tiny chip. Open the hood of your car and you'll find little black boxes filled with chips, and not much you can fix yourself. The tuning knob on your radio has disappeared, and even musical instruments have changed from mechanical contraptions to synthesized sound. And today's huge airliners rely on "fly-by-wire" controls—fewer moving parts and fewer failures.

So, just as I joined that elite group of dancing people, I want you to join a select group: Only one person in ten thousand knows Jack Kilby's name—so join that group and celebrate the achievement of Jack Kilby whose brilliant invention spawned the billions of microchips that run our watches, cars, jets, computers, calculators, and .... metronomes.

Baseball Stats

THIS WEEK THE ALL STAR GAME MARKS mid-season for baseball. So, today I look at "America's Pastime" with my engineer's eye.

With the touch of a button they track an entire game in real time. These computers compile statistics too. Not the simple ones of years ago, but more arcane measures like "scorability", which measures a team's efficiency at scoring runs; or a player's "seasonal notation," a measure that takes all of a player's stats, and creates a virtual season that represents a year's worth of their "average" stats over a career.

And for these new players being in the big leagues isn't as simple as just swinging a bat. The San Francisco Giants have a "video coaching system" to help players hit the ball. It features six DVD drives to archive over 4,000 hours, that is, an entire season's worth of pitcher-batter match-ups. Coaches search the database on any computer in the Giant's internal web and view an at-bat from four camera angles.

Of course, even that bat isn't a low tech thing nowadays. Engineers study the collusion between a pitch and the aluminum bats used by college teams. They've isolated a "trampoline effect." That is, the thin wall compresses during the collusion with the ball and springs back. So, manufacturers now make aluminum bats with walls of just the right thickness to give that extra oomph. The Major Leagues don't allow aluminum bats, but their wooden bats are also heavily researched. And once the players swing the technology isn't over. Boston's Fenway Park has an Umpire Information System. It uses pairs of cameras to track pitches. One set in the rafters above the first base line, and one above the third base line. Two additional cameras in the dugouts take a snapshot of the batter just after the pitch. A computer uses these images to determine, within half an inch, the path of the ball.

The system doesn't replace umpires or even second-guess them, it's used to help them refine their calls. After the game, it even generates a CD, which a ref can study in private. If you're a baseball purists I'm sure all this sounds like too much. Take heart, though, Major League Baseball still prohibits computers in the dugouts. Manager's instincts and hunches still control the game, although they do allow managers to consult pages of statics generated by computers.

Waterless Urinals

I EXAMINE RESTROOMS WHEREVER I travel because their design captures some essence of a society since every strata of humanity needs to use one. This fact makes a toilet a very political thing.

For example, last week I used a rogue, in fact nearly illegal toilet: A waterless urinal at Chicago's O'Hare airport. While good enough for the Taj Mahal and Disneyworld, they aren't up to par for the State of Illinois.

The urinal's plastic cartridge at its bottom contains a thin layer of oil, which forms a trap for odors—so no need for water for flushing. These urinals offer great advantages over the standard units: They use less water, take less plumbing and maintenance, and they're hard for vandals to clog with paper towels.

Sounds like a no brainer for most businesses to install waterless urinals, yet Illinois's plumbing code prohibits them. To install them in Chicago the City got an exemption that allowed a 60-day test.

Why the resistance? Well, as I said a toilet becomes the nexus for all manner of social forces and human behaviors. Plumbers Unions, for example, oppose the new urinals. No water means less plumbing, which means, of course, less work. So, they've mounted a mighty protest claiming that the new urinals smell. They also assert that germs grow on the plastic insert, suggesting that without a steady stream of water bacteria will become glued to the toilet and contaminate the world.

But it isn't the Unions alone that prevent installation of the new waterless urinals. Human nature plays its part.

Imagine if you were in charge of constructing a large building, which needed water closets on each floor. Would you choose the flush urinals that have worked for years, or some kind of new-fangled technology? Human nature pushes toward the former.

Not surprisingly, then, what's been accelerating the waterless urinal's acceptance so far has been mother nature. The first surge in sales occurred in the mid-1990s when harder than expected drought hit the west. Even so, it isn't really nature, it's politics that drives the adoption of the new toilets. The drought forced politicians dealing with water shortages to allow waterless urinals. In fact, right now their sales double every year.

So, even if on some rational level a waterless urinal makes sense it shouldn't surprise us that adoption depends on all sorts of factors. Aristotle got it right over 2,000 years ago when he wrote: "Man is by nature a political animal." That in turn makes humankind's tools and conveniences, like urinals, subject to the whims of politics.

Dvorak Keyboard

I'M SURPRISED AT THE NUMBER of listeners who've asked me to tell the story of QWERTY . That word refers to the arrangement of letters on a typical keyboard: Look at the upper left hand row and you'll see the keys Q-W-E-R-T-Y. You see, listeners want me to talk about it because it supposedly proves the pure irrationality of the technological marketplace.

The designers of the QWERTY keyboard, the legend goes, placed the keys randomly to slow down typists. And once adopted no other keyboard could take over. The prime piece of evidence for this is an alternative keyboard called the Dvorak layout. It is vastly superior, but alas was squashed by the inferior QWERTY keyboard, only, though, because the QWERTY keyboard entered the market earlier.

Now, here's the true story and its moral.

August Dvorak, a professor of educational psychology, redesigned the keyboard using the then current notions of time-motions studies. A true believer in a purely scientific approach to life he urged his pupils to "Make yourself efficient and up-to-date, wherever possible, by the use of available machines." In short," he said use machines to "be civilized."

To Dvorak, it appears, nothing proved a greater threat to civilization than the nasty QWERTY keyboard. He found it to be, in his own words, "so destructive that an improved arrangement is a modern imperative." So, he designed a keyboard that minimized fatigue and was easier to learn.

Or, so he claimed. Alas, for the benefit of the Western World, the evidence is scant.

Most of the evidence of the superiority of the Dvorak keyboard rests on studies done by the Navy during World War II. That is, until one

\

looks more closely. The report's author was one Lieutenant August Dvorak—the Navy's leading time-motion expert.

In modern times researchers detected either no advantage or a slight increase of four percent, at best, in typing speed. In fact, one researcher concluded "[d]o not waste time rearranging the letter[s]" on a keyboard.

Oddly, many take the moral of the Dvorak keyboard's failure to be that the technological marketplace doesn't change, that we'll live with an inept device simply because the manufacturer got to the market first.

Yet, the real lesson lies in the new keyboard's tiny, at best, improvement over the QWERTY keyboard: To dislodge an existing technology requires a significant change in performance, and likely an increase in functionality.

The marketplace does indeed change: Just ask the manufacturers of vinyl records or cassette tape, they've long been displaced by CDs and MP3 players. These new devices do the job of their predecessors, but —and here's the key—they also offer vastly different capabilities. You would never, after all, use records or cassette tapes to share music via the internet.

NOTE ON THE TYPE

This book is set in Adobe Caslon, a superb digital rending of the typeface designed by the great English type designer William Caslon (1692-1766). Type designer Carol Twombly (1959-) has captured the unique character and charm of this 18TH century typeface, used in the *Declaration of Independence*. Caslon began his career as an engraver of pistols and muskets, eventually moving into letter founding, which made him famous in his lifetime. He produced every letter carefully over a period of twenty years, cutting each by hand. This resulted in idiosyncratic, but highly legible type. Some designers take a dim view of this individually, but despite this professional disdain the typeface remains one of the best loved.

Made in the USA
Monee, IL
15 February 2021